An Introduction to Color Forms of the Domestic Fowl

A Look at Color Varieties and How They Are Made

by

Brian Reeder

Bloomington, IN

Milton Keynes, UK

AuthorHouse™
1663 Liberty Drive, Suite 200
Bloomington, IN 47403
www.authorhouse.com
Phone: 1-800-839-8640

AuthorHouse™ UK Ltd.
500 Avebury Boulevard
Central Milton Keynes, MK9 2BE
www.authorhouse.co.uk
Phone: 08001974150

First published by AuthorHouse 2/8/2006

ISBN: 1-4259-0421-1 (sc)

Printed in the United States of America
Bloomington, Indiana

This book is printed on acid-free paper.

My thanks to my grandparents who encouraged and supported my interest in poultry and to all the creative ancestors who left us their legacy of living art; the many color forms of the domestic fowl. It is to their memory and achievements that I dedicate this book. My best wishes to everyone in all their projects.

Table of Contents

INTRODUCTION

In this volume I will be looking at the interaction of the e-alleles with other color/pattern-effecting genes, and how those combine to make varieties. I hope to present a body of language which can help communication between all sectors and countries; a nomenclature which is consistent and based upon the terms for the various genes. In other words, a "universal naming system".

I will then make an attempt to list as many actual recognized varieties as possible, list the genes that combine to make that phenotype and then offer the new universal name for that variety. In so doing, I hope to offer to the hobby, around the world, a consistent series of names, based on a logical system which is rooted in the names of the genes and specifically based on the e-alleles upon which that variety occurs. Only a few instances may not fall into this system, due to the ability to produce that phenotype on more than one e-allele, with equally fine results. These forms will be noted as such.

In this way, I hope to foster communication between the breeders of different breeds, varieties or strains as well as breeders from around the world and help begin the process of understanding the nature of the color and pattern varieties of poultry, as seen in the hobby.

Brian Reeder

THE GENETICS OF COLOR AND PATTERN

The genes of color and pattern are a complex subject, and everything about this subject is not yet known. However, enough is known that we can deal with this subject in an intelligent manner. While many different papers have been written on the various genes in this group, and many people know, use or deal with one or two small sections of this information as regards the specific variety that is being raised, very few have been able to put this together in a cohesive whole and understand the structure of these varieties, and how their genes interact. As well, relatively few have been able to establish the structure upon which all the patterns or varieties are built.

There is a specific structure that can be followed to determine the genotype of a given variety. This goes as follows; e-allele, s-allele, and then the other additive genes that finish out the variety. That is also the way the genes are to be listed. When those gene notations are translated into a name, they are listed in reverse order. Here is an example. Let us say that we are talking about the silver phoenix, as standardized for color pattern in America. It's gene notations are listed as e+/

e+ S/S. When we translate this into the color variety name, we call it silver duckwing (S = silver and e+ = duckwing). We will deal with this in much greater detail in a later chapter, but for now, it lays the order of the gene notations, and the order of those genes when listed as a variety name.

To get started, we must first look at the pigment that produces colors seen in poultry. This pigment is melanin, and is divided into two forms; pheomelanin and eumelanin.

Eumelanin and Pheomelanin

These pigments are both forms of melanin. Eumelanin is the "black" pigment, while the pheomelanins are the "red" pigment. They produce different "colors" through enhancements or dilutions, producing all the shades known in poultry, in recombinant with these other genes. We will discuss both pheomelanin and eumelanin below.

Pheomelanin

Pheomelanin is characterized in the jungle fowl as red to orange to salmon. In it's wild form, it is a "warm tone". With dilution it can become gold or silver. When it becomes silver, it appears to be "white", if no "red enhancers" are present. There are two forms of pheomelanin in the domestic fowl; Autosomal pheomelanin and sex-linked pheomelanin. In the wild type e-allele e+, the salmon breast on the hen and the darker reddish brown pheomelanic area seen in the wild type males shoulder is a different form of pheomelanin from the much better known sex-linked pheomelanin (S/s+). I have named this factor Autosomal pheomelanin, in reference to it's gene action and it's difference from sex-linked pheomelanin.

Autosomal Pheomelanin

This form of pheomelanin is not on the s-allele and is not affected by S (silver). I notate the gene Ap. Note the first letter is capitalized, as this gene is dominant. This form of pheomelanin is also found in the entire body of eWh wheaten e-allele hens and is actually present as Ap or ap+ in all of the e-allele mutations (more on this later). The mutation at the Ap allele is ap+, which is the absence of the Ap gene. This then allows for dilution effects. Ap is a dominant gene with variable expression and incomplete dominance in the heterozygous state (Ap/ap+). Jeffrey called this second pheomelanic form "autosomal red". I would suggest that Autosomal pheomelanin (Ap and ap+) is more appropriate nomenclature.

The strongest effect of Ap is seen in the body. While Ap does have some effect on all pheomelanic areas, it's greatest effect is seen in the body, with the fullest saturation in the breast (though it is masked by eumelanin in the male), the back and the shoulder (wing bow). The effect of Ap in the hackle, saddle (cushion in the female) and the main wing feathers is much less than in the previously noted body areas. Sex-linked pheomelanin dominates these areas. There will be more on this gene (Ap) and it's opposite mutation (ap+) throughout the following text, where we will look at interactions and recombinants of Ap-ap+ and S-s+ with the e-alleles and the modifier genes.

Sex-linked Pheomelanin

Sex-linked pheomelanin is the commonly known form of pheomelanin, S and s+. This gene is sex-linked, being on the z-chromosome. Thus females can only have one dose (as they only have one copy of the z-chromosome) while males have two copies. This allele has the strongest effect on the areas of the hackle, saddle (cushion) and main wing feathers (the "duckwing" triangle of the lower wing). We will discuss the recombinant effects of S-s+ with other genes under each of those sections. This allele has been studied heavily, due to it's commercial applications in down-sexing of day old chicks, and the value of this should not be underestimated in any breeding scheme.

Eumelanin

Eumelanin in the jungle fowl is black. In the hens, however, there is a further interaction between the pheomelanins and eumelanin that produces intermediate shades called "browns". Further, eumelanin can be diluted to make blue gray, chocolate brown, light cream brown, lavender gray or white. Eumelanin can be extended into normally pheomelanic areas to create "darkened" varieties, ranging up to solid black, which can then also be diluted into various lighter shades, all the way down to solid white.

All the colors of poultry are made up of these two pigments (three forms, including both forms of pheomelanin). It is the interactions or the functions of other genes modifying these pigments that creates the range of shades and tones seen in poultry.

It is the e-allele that actually determines where the

pigments are distributed on the bird, for each sex. All the e-alleles show some level of sexual dimorphism, thus we can conjecture that this locus is influenced by or has some effect on, hormone levels and/or hormone production.

An understanding of the e-allele is necessary. In order to understand the basis of the varieties, one must understand the e-allele. It is the base upon which the varieties are based. I thus use the e-allele as the basis of all variety naming in the system I will present throughout this volume. With this system, the varietal makeup on the genetic level is directly expressed. I believe that a world wide basis for nomenclature is essential for all efforts in the future, as we become and increasingly global community.

It is thus to the e-allele that we turn next. We will present a very basic outline of the major e-alleles seen in poultry varieties, to establish the basis for understanding all the varieties that will be presented later.

THE E-LOCUS

Now that we have established that we are only working with two pigments, we need to look at the loci that distribute these two pigments into certain specific patterns, all of which are sexually dimorphic. The genes that do this are the various mutations at the e-locus.

These mutations are called alleles. They distribute pheomelanin and eumelanin into very specific areas of the male and female in all of the e-alleles. These e-locus alleles are the basis of all the patterns and varieties, for to make the patterns or varieties correctly in both sexes, they must be on the correct e-allele. There are a few patterns which will work well on several e-alleles, but generally the pattern is dependant on the correct e-allele. Thus the e-alleles are the basis of all patterns and the place where we must begin in understanding varieties.

There are five e-alleles that are found in the patterns and varieties of the vast majority of chickens. These are e+ (wild type, known as duckwing), E (extended black), ER (birchen), eWh (wheaten), and eb (brown).

We will start at the beginning, with e+ (wild type or duckwing) as is found in the wild red jungle fowl.

Wild type or Duckwing (e+)

This allele is the "original" e-allele and is the pattern seen in the wild red jungle fowl. It is called "duckwing", due to the patterning of the males wing being thought to resemble a mallard duck's wing. Males of eb (brown) and eWh (wheaten) are often vernacularly referred to as "duckwing", but this is only a reference to the male pattern and only e+ has been named duckwing by the geneticists.

In this allele we see sexual dimorphism, the males and females being different in distribution of eumelanin and pheomelanin. The chick down of e+ (when not effected by other dilution or enhancement genes) is striped or "chipmunk-like" as some refer to it. It is referred to as agouti and is a pattern seen in many wild animals. The pattern of chick down has two prominent stripes running down the back, from the sides of the beak, back across the eyes, down the neck, over the back and to the rump. The underside of the chick tends to be lighter colored, as is the middle section between the stripes. In the wild (s+) red form the chick is variations of warm browns and beiges. When S is added they become silvery grey, cinnamon and black. Other genes can modify the appearance of this allele's down.

The adult plumage of the male is divided into specific sections of eumelanin and pheomelanin, with both forms of pheomelanin present. The hackle, saddle, and main wing feathers are all sex-linked pheomelanic (red, gold or silver depending on the variations of pheomelanin). The "shoulder" and back are Autosomal pheomelanic and are usually darker than the sex-linked pheo areas in the unmodified wild type. The rest of the body, breast, remaining wing sections, thighs, underfluff, tail and sickles are eumelanic (black or dilutions thereof).

The plumage of the female is very different from the male. Her back, shoulders, cushion (female equivalent of saddle feathers in the male), lesser sickles, wings and underfluff are made up from a combination of eumelanin and both forms of pheomelanin that appears warm brown in the wild type when s+ is present (red duckwing). These areas have eumelanic "stippling", that is much like broken up partial lacing, black sandy speckles and pattern in the individual feathers. This is most notable in the back and wings of the females. The hackles are orange with a wide black stripe down the middle of the feather. The breast of the e+ hen is Autosomal pheomelanic, and it is not affected by the S/s+ allele, except when ap+/ap+ is present, as it remains in reddish tones even when S is present. The breast is salmon, though heterozygosity or absence of Ap (Ap/ap+ or ap+/ap+) will create variation in the tone of the breast, as will certain diluters or intensifiers, yet the basic distribution of stippling and eumelanin/pheomelanic blending on the back and pheomelanic saturation of the breast with no eumelanic stippling or blending will remain consistent. Each feather of the breast tends to have a lighter area on the end of the feather much like a slightly lighter pheomelanic lace. This "salmon breast" is a key to identifying e+ birds, as even when other genes are present which may obscure the allele, the breast of the e+ hen always remains pheomelanic. Any hen that is e+ will be obvious by the variations of colors and patterns that divide the breast apart from the rest of the pattern/color. Even in extreme modification by many genes, the breast will be a slightly different shade and will have some obvious differentiation from the rest of the bird. Autosomal pheomelanin also has some presence in the back of the e+

hen and is especially strongly concentrated in the "shoulder" and mid back of the hen involved in this recombinant phenotype.

WHEATEN – eWh

Wheaten is a sexually dimorphic mutation. It seems to be an extension of pheomelanin and a reduction of eumelanin, especially in the female. Basically the breast coloring of the e+ hen is extended to the entire body of the hen in wheaten. For all intent and purpose, the rooster of eWh remains unchanged from the roosters of e+. Thus it is the hen or the chick down which we must look to in determining whether we are dealing with e+ or eWh.

The chick down of eWh tends to be a solid white or yellow, although some other genes (such as Pg or heterozygosity at the e-allele, for instance) can add some pattern to the down, but in general, the eWh chicks are solid yellow or solid white (additions of certain pheomelanin extenders such as Co and Db can change the shade of the down, more on that under those topics). This is quite a departure from the e+ chick with it's agouti pattern. The hen of the wheaten has a solid wheat colored body, varying in shade from nearly off white to dark cinnamon depending on the presence/absence of Ap/ap+ and Mh/mh+, much as does the breast of the e+ hen. Wheaten hens which are very clean and light on back and breast are ap+/ap+ mh+/mh+, and these will in some ways superficially resemble a silver Columbian phenotype, except that the hackles will be orange to gold. Variation within the tones indicate the recombinations of Ap-ap+ and/or Mh-mh+ with this allele. When in recombinant with Ap and/or Mh, the eWh hen can show a darker back and lighter breast, which can seem to mimic e+ hens to the untrained eye. However, the back will be a warm dark cinnamon while the breast is a lighter shade, ranging from salmon to near white. The back will show none of the stippling associated with e+ however. The main tail feathers are eumelanic as are the main wing feathers. The hackle is usually a solid eumelanic color, but some will have a small bit of striping. The rooster of the eWh allele is identical to the roosters of the e+ allele.

Wheaten seems to be an "extension of pheomelanin", as per the hen and the lightened down of the chicks, or could be referred to as a "restriction of eumelanin". Either way of describing this allele is correct. Wheaten has been noted to lighten the melanin of the shank (leg), but there are some stains of wheaten which have dark shanks. The interactions that make the dark legged wheaten are not clear at this time, the darkening factor not having been studied by myself or any researchers who's work I have studied.

There is said to also be a recessive form of wheaten that was named ey, but this may not in truth exist. Clive Carefoot, a British poultry research and show breeder, has determined that eWh is dominant in the absence of melanizers (such as Melanotic or charcoal), but becomes recessive in the presence of said melanizers. I have tested this theory and found it to work.

Brown – eb

Brown, eb, is a sexually dimorphic e-allele, and is basically a "darkening" of the hen, where the entire body of the hen is like the back of the e+ hen. There is no salmon breast in the eb hen, this being the major trait that separates this allele from the e+ allele. Thus this allele is an eumelanically extended mutation. While Ap is present in many varieties of this allele, the normally Ap breast of the e+ hen has been covered by the patterned and blended eumelanization as seen in the e+ hen's back. The chicks vary depending on other genes and this allele may have some of the greatest variations in chick down seen. The basic eb s+ chick is quite like the e+ s+ chick, but they are more diffused in their striping down the back.

The eb hen is throughout the body a combination of eumelanin and pheomelanin. Her whole body is much like the back of the e+ hen. It is a bit darker with the stippling being heavier and the patterning being a bit darker and heavier, giving a partial double or triple laced effect. Without a close inspection, the eb hen looks very much like the e+ hen without the salmon breast. Her main tail feathers are eumelanic and her hackle is pheomelanic with a wide black stripe in the center of each hackle feather. The rooster looks like the e+ or eWh rooster, except he will tend to have dark striping in the hackle and saddle. Some e+ or eWh roosters can have this striping too (in recombinant with melanizers), while others are clean of striping in hackle and saddle, with eWh tending to be the cleanest due to the eumelanic restriction and pheomelanic extension seen in eWh. It is very rare to see an eb rooster without the hackle and saddle striping (except in certain recombinants, such as Db). Of all the e-alleles, eb is the most recessive, being recessive to all the other alleles. It is however the case that eb is dominant to eWh when, and only when, there are melanizers present, as was discussed in the section on eWh. A great many varieties can be made on eb and the patterned varieties are especially effective on this allele.

Birchen – ER

We now move into the area of the fully eumelanically extended versions of the e-allele mutations. E and ER are the least sexually dimorphic of the alleles at the e locus. ER produces hens that look very similar to the roosters, with only the "sex feathering" (saddle and shoulder in males) being different from the hens eumelanic/pheomelanic distribution. The chicks are generally solid black. With other genes, this down coloring can be greatly altered, perhaps more than any of the other alleles (perhaps only eb exceeds ER, but then again, they could tie). ER down allows the visual expression of Co and Db. This differentiates E from ER, as E down will not allow the visual expression of Co or Db. Some ER downs are nearly solid black with a tiny amount of white on the underside of the chin, but generally, unless other genes are affecting it, the ER down will be solid black to black with a dark brownish red shading on the top of the head.

The hen of the ER allele is a completely eumelanic bird with pheomelanin expressed only in the hackle and as lacing on the breast. The hackle is pheomelanic with a wide eumelanic stripe down the middle of each hackle feather. The breast is eumelanic with a lacing at the edge of the feather that is pheomelanic. In my experience, this allele can have either Ap or ap+, which will have little effect on the unmodified E~R phenotype. The rooster at first glance appears quite like the rest of the roosters described in the other e-alleles we have discussed. However, there are some distinct differences. First, his entire wing is eumelanic, with only the shoulder being pheomelanic. This is then referred to as "crowing", as it is said to resemble the wing of the red winged black bird, a crow type bird. The ER rooster is eumelanic in all areas except the hackle, saddle and shoulder, with the breast showing the same pheomelanic lacing as the hen. The hackle and saddle has the wide eumelanic striping and the breast feathering has an edge lacing of pheomelanin, which appears to be sex-linked pheomelanin in both sexes. This allele allows the expression of Co, Db, etc., so there are several very surprising and beautiful patterned birds which can be made on ER. This is the greatest difference between ER and E.

ER also darkens the shanks, a deep slate melanization covers the shanks. It is possible to produce ER based birds without the heavily melanized legs, but these tend to be seen on the "fayoumi" type of birchen and/or in birds which have one dose of another e-allele along with the dominant E~R, and often with the Id (dermal inhibitor) gene as well.

EXTENDED BLACK - E

Many folks assume that E makes a solid black bird, but in reality, E can exist without any modification for melanic extension and look almost identical to ER, except for a great deal less breast lacing, if any. I refer to these by their form of S-allele phenotype with the word "black". For instance, red black, orange black, golden black, light golden black, silver black, mahogany silver black, etc.

There is very little difference in E and ER in phenotypic expression. The downs are more markedly different than is the adult forms with the same melanic mutations. The E down is typically darker on top and lighter underneath. Generally black above and white to yellow below, though the shades will vary with modification of the eumelanin. This allele is known to produce shorter down, which can cause some level of difficulty in hatching. E is the only e-allele which has so far shown a deleterious plieotropic effect.

For making a black chicken, eb, ER or E are all equally usable. E is not the only allele upon which a solid black phenotype in plumage can be created. It in fact may be the worst one to make this phenotype on, due to the short down effect of the mutation. ER or eb may be very much preferable for creating solid black and it's allied varieties.

E causes a black shank coloring in most instances, though it may not be very solid at first and may only darken over the first few months. Some researchers in the past have claimed that Id (dermal inhibitor) can have no effect on the shanks of the E fowl, yet this does not seem to hold true in all instances. As for ER, heterozygosity at the e-allele can certainly be a factor in Id expression on E.

To make a solid black bird, one must add melanizers, to extend melanin into the normally pheomelanic areas of eb, ER or E. This will be discussed under the discussion of the melanizers. Some have put forth that E alone makes a solid black bird, yet this is not the case. There is no single gene which makes a self black phenotype. In fact, self birds (except for recessive white) are some of the most complex varieties to create.

The order of dominance in the e-allele is ***E->ER->eWh*** (without melanizers present***)->e+->eb->eWh*** (when melanizers are present).

Let us now consider what we have learned in the above discussion of melanin and the e-alleles. All pigment in all chickens is melanin, either eumelanin or pheomelanin, with the pheomelanin existing in two distinct forms. When notating this phenomena, let us consider the basic genotype of the "wild type". In notating this basic color form, we must consider the e-allele and the forms of pheomelanin. Thus wild type is notated as e+ (duckwing e-allele) s+ (gold or "red") and Ap (presence of Autosomal pheomelanin. The homozygote is e+/e+ s+/s+ Ap/Ap (for males) and e+/e+ s+/w Ap/Ap (for females, as she only has one z chromosome, that being the sex chromosome, of which the male has two.) This is then called red duckwing.

This process would working the same manner for any of the other e-alleles. As one other example, let us look at the eWh allele. The most basic version of eWh would be the red wheaten. This would be eWh s+ Ap. In males it is eWh/eWh s+/s+ Ap/Ap and the female would be eWh/eWh s+/w Ap/Ap. It would be called red wheaten. This process would be repeated for any of the e-alleles, citing as many genes as one is aware of, either based on phenotypic identification or testmating confirmation.

Now we know that all the colors in poultry are caused by two pigments in modification of tone and saturation and that the e-allele is responsible for the base distribution of those two basic forms of pigment; eumelanin and pheomelanin. In other words, the e-allele mutation determines what the ratio of eumelanin to pheomelanins is and where those are distributed, especially so in the hen. Areas of color can be Eumelanin, Autosomal Pheomelanin, and/or Sex-linked Pheomelanin as well as the layering effect of Sex-linked Pheomelanin and/or Autosomal pheomelanin with Eumelanin as seen in the bodies of eb and the backs of e+ hens. These are the four direct fronts upon which the fowl is "colored", thus creating a wide range of phenotypic possibility depending on the mutations present. The extent of saturation of pigment seen in any of the above four examples would depend upon the genes that I refer to as Diluters, Intensifiers, and/or Extenders.

Next we will list the genes currently known to effect phenotype, with a short definition of each.

GENE LIST

This section will deal with all the genes which effect the pigments; eumelanin and pheomelanin (both sex-linked and autosomal). It is the various effects of these gene, on a given e-allele that creates all the varieties.

I will be listing the various feather color/pattern genes in groups of genes producing similar phenotypes. It is not be implied that these genes are related on a molecular level.

The groups of genes are as follows.

1. ***Diluters which have a greater influence on pheomelanin*** - which turn wild type s+ and Ap to lighter tones including "gold", "buff", "cream" to "clean white", Silver.
2. ***Diluters with a greater effect on eumelanin*** - which turn eumelanin (black) pigment to lighter tones such as "gray" (Blue), "brown" (Dun, chocolate) "Cream" (Splash, Homozygous Dun) to white (Dominant white).
3. ***Diluters with an equal effect on both forms of melanin*** - One diluter dilutes both forms of melanin; eumelanin and pheomelanin (both s-allele and Ap). Eumelanin turns to a

purplish lavender tone called lavender, while pheomelanin is turned to varying levels of straw, to yellow to light pale peach/orange ("orange" sherbet color), due to the effects of the recessive gene lavender.

4. ***Intensifiers of pheomelanin*** - Mahogany, which makes red darker on either s+ (when non-diluted) or Ap.

5. ***Extenders of Pheomelanin*** - which cause pheomelanin to enter into the typically eumelanic areas, most notably turning the eumelanic areas of breast and body to pheomelanin.

6. ***Extenders of eumelanin*** - which extend eumelanic pigment into typically pheomelanic areas. These genes tend to darken birds, increasing eumelanin, most notably in the sex-feathered areas of male birds.

7. ***Genes removing all pigment*** - melanic production disruptors. These create totally white birds, by elimination of all pigment. Some forms are more effective than others.

8. ***Feather-pattern modifier genes*** - these change how the eumelanin (or it's absence) is distributed on the feather. Pg is a notable gene in this group for having a wide interaction group and actually being responsible for several of the patterns which are well known and much desired in poultry varieties.

It is the actions and interactions of these genes which manifests as all the varieties known in fancy poultry.

Diluters which have a greater influence on pheomelanin

1. **S - Sex-linked Silver** - S is a dominant allele, whose alternate is s+ (red/orange/gold - wild type). S has a strong

effect on sex linked pheomelanin, but a much reduced effect on Ap.

2. **ap+ - Absence of Autosomal pheomelanin** removes the Autosomal pheomelanin. Appears to be a recessive, with wild-type Ap being dominant, expressing in the heterozygous state.

3. **Di - Dilute** is most effective on the s-allele, with little effect on Ap. It is an autosomal dominant. It turns s+ to lighter shade and assists S in making a cleaner, lighter tone.

4. **Cb - Champagne blonde**. A common gene in the buff phenotype. It may influence other forms but it is not recorded to date. Autosomal dominant, it evens" the tone of both forms of pheomelanin allowing for an even shade of pheomelanin in all parts of the feathers. Also considered a diluter of pheomelanin, turning s+ to lighter shades.

5. **ig - inhibitor of gold**. An autosomal recessive, which dilutes s+ to a light cream tone in homozygotes. Not utilized in many varieties, it is generally found in the "cream light brown" Old Dutch bantam.

Diluters which have a greater Influence on Eumelanin

1. **I - Dominant white** - a diluter of eumelanin, which turns black to white. More effective in some backgrounds than others and is considered a "leaky" white. Thus there may be several genes present to enhance the whitening effect of I. Dominant white has a mild dilution effect on s+, with little effect on Ap.

2. **Bl - Blue** - A diluter that is an autosomal dominate. In it's heterozygous form it creates the Blue coloring, a varying

grayish-blue. In it's homozygous state it creates "Splash" which is a highly variable and creates a whitish phenotype.
3. **Id- Dun** - an autosomal dominant, which behaves in a identical manner to Blue, but manifests in shades of Brown.
4. **"rc" - recessive chocolate** - A gene which is rare in the US and only documented amongst Serama. There is no abbreviation for this gene at this time. I thus use "rc" (recessive chocolate), when abbreviating this gene. It is an autosomal recessive which turns black to medium brown (almost identical to heterozygous Dun) in it's homozygous state.

Diluters with an equal effect on both forms of melanin

1. **lavender - lav** - a gene which dilutes both eumelanin and pheomelanin. The effect on eumelanin is to turn black to a purplish-gray, while the effect on pheomelanin varies from a pale straw/cream to a peach-orange. This gene is an autosomal recessive and has no effect in the heterozygous state.

Intensifiers of Pheomelanin

1. **Mh - Mahogany** - This gene intensifies red, turning it to a deep dark, blood red. The strongest effect of Mh is on Ap, but Mh can also have a strong effect on s+, IF Dilute (Di) is not present. Mh is an autosomal dominant.

Extenders of Pheomelanin

1. **Co - Columbian** - This gene extends pheomelanin into the breast and body of both sexes. Co is an autosomal

dominant. The heterozygote shows variable penetrance of the gene, but homozygotes tend to show very high penetrance. Any variation of pheomelanin can be seen, though it has never been conclusively shown that Co can be present in the extremely dark red forms of pheomelanin. Co seems to dilute pheomelanin as well as extend it, though it's primary function is as an extender of pheomelanin. It should be noted that Co is more effective, both as a homozygote and a heterozygote, at extending pheomelanin than is Db.

2. Db - Dark Brown - "ginger" - This gene is much like and easily confused with Co. However, there are several distinctions between the two. First, the interactions of Co and Db with Pg are different (more on this below). Second, Db and Co create different "shades" in recombinant with s+. Third, Co is more effective than Db in the extension of pheomelanin, with Db showing a wider range of penetrance than Co. Db is commonly called "ginger" in the hobby, and is a varietal trait of "ginger red" games, amongst others. Db is an autosomal dominant. The heterozygote can show wide variation and thus has highly variable penetrance.

Extenders of Eumelanin

1. Ml - Melanotic - Melanotic is a commonly encountered gene. It is essential in the making of several patterns when recombined with Pg and other factors. Ml is an autosomal dominant and shows variable penetrance in the heterozygous state. Many genes can interact with Ml to make varying phenotypes. Ml has the strongest effect on the hackle feathering in both sexes on the top of the head and in the upper half of the hackles. Ml also interacts with any pattern present to "darken" and eumelanization these areas, be they stippling, Pg, etc. Ml is most effective in creating patterns. I have not seen any evidence that Ml in recombinant with E or ER alone can make a solid black bird, as Ml does not seem to be very effective in eumelanizing the lower hackles. It seems that Ml and/or E/ER requires further melanizers to achieve a solid black phenotype.

2. cha - charcoal - This is a recessive eumelanin extender which has been described in UK. I have never worked with a known example of this gene, but I have worked with birds whose eumelanic genes seem to act in a similar fashion. I however can not confirm nor deny that this was cha. This gene is an autosomal recessive. Perhaps many of the commonly encountered "recessive" melanizers are cha. Only a test mating of confirmed cha to these other forms will confirm or deny what these other genes are or are not.

3. "rb" - recessive black - This is not a named gene, but rather is a catch-all for these "recessive melanizers" seen in many varieties. Some may be cha, others could represent "ebonies" as described by Jeffrey, while others may have never been named in any respect. I use this symbol as a catch all symbol, for a class of eumelanin extenders that are not well documented in the literature, but which, non-the-less, are present and important factors in many varieties. My work with these genes shows the following; they are recessive in males, yet dominant in females, indicating a sex-expression of dominance. These genes have the strongest effect on the lower hackles of males and females. In most instances, any one factor for eumelanization did not make a fully black phenotype in recombinant on E or ER. I believe that generally, two eumelanizers are present in most fully eumelanically extended varieties (i.e., black or black varieties {blue, dun, etc}) Thus please be aware, that when "rb" is shown in a genotype, this is to represent the general term "recessive black", which is not a named, documented, specific gene, but rather, the confirmed presence of this group of factors. There is more than one form of this "recessive black". In outcrossing many black lines, I have seen instances of confirmed "non-melanotic" (ml+/ml+) lines, which produced 100% solid black birds in both sexes when bred pure, but these black lines when outcrossed to each other produced 100% of males with some level of pheomelanin showing through in hackles, saddles and shoulder. This then indicates that there were at least two non-allelic genes at work, and thus non-compatible, and not producing the solid black phenotype. One of the most interesting secondary effects I have repeatedly noted from at least one gene in the "rb" group, is the ability to intensify the depth of red pheomelanic pigment. This is observed in the very dark Rhode Island Red seen in exhibition. These birds clearly show true black, heavy striping in the lower hackles and more black in the main wing and main tail feathers, than in other red varieties. I think this effect is a recombinant of Mh with absence of Dilute (di+). Another pattern is that in those eWh based varieties which have strongly black main tails and have some small amount of striping in the hackle such as New Hampshires, the recessive black factor is recombinant with Di, to create the lighter coloring and lesser amount of Black seen in all parts. I do not know if all forms of recessive black "rb", would show such an interaction pattern.

Genes removing all pigment

1. recessive white - First, I must mention that there are two genes which create a recessive white phenotype. One is more effective than the other as the less effective form has less effect on autosomal pheomelanin, allowing it to show up on eWh and e+ background genomes.

One type has been described in the literature and named "c", which is the proper recessive white. While there are two genes here, which are non-allelic, I will use the gene designation "c/c" for any line known to be recessive white, as the second form has not been named. Thus anywhere that the abbreviation "c/c" is found, this could refer to either form of "recessive white".

Both forms are autosomal recessives. The heterozygote shows no penetrance, and is not visually identifiable, while the homozygote shows total penetrance.

Feather-pattern modifier genes

1. Pg - Pattern gene - This gene effects the distribution of eumelanin in the feather and is highly malleable. Many patterns can be achieved depending on the other genes present which interact with Pg. Pg is an autosomal dominant, with heterozygotes showing variable expression of the trait as expressed through any interaction group. The chart below shows the interaction groups for Pg. Pattern gene varieties are seen commonly on two of the e-alleles; eb and ER. The other forms are much rarer.

e-allele	interaction genes	variety
e+	Pg Db	Red Quill (*as in O.E.G.*)

e+	Pg Ml Db	Spangled (*as in Red Shoulder Yokohama*)
eWh	Pg Ml	"Dark" (*as in Red caps and O.E. Pheasant fowl and perhaps some lines of Cornish, Malay, Shamo/Aseel)*
eb	Pg	Partridge (*in cochin*) Penciled (*as in Brahma or Wyandotte*)
eb	Pg Db	Autosomal Barred (*"penciled" as in penciled Hamburg*)
eb	Pg Ml	Double laced (*as in Cornish and Barnevelder*)
eb	Pg Ml Co	Single laced (*as in Wyandotte*)
eb	Pg Ml Db	Spangled (*as in Golden Spangled Hamburg*)
ER	Pg Db	Autosomal barred (*as in Campine*)
ER	Pg Ml Db	Spangled (*as in Silver Spangled Hamburg*)
ER	Pg Ml Db Co	Single laced (*as in Polish and Seabright*)
E	Pg Ml Co Bl	Laced Blue (*as in Blue Andalusian*)

The pattern gene interactions are very interesting and results in a great number of very ornamental and popular varieties.

2. B - sex-linked Barring - This is an interesting gene, which though much studied is not well understood. This gene has a tight linkage with Id (Inhibitor of dermal melanin). This linkage pattern is ten to fourteen centrimorgans in distance, thus a barred bird with dark shanks is rare, but they do occur. Further research needs to be done to determine the full nature of B. B is said to be an eumelanin diluter. I would classify it as a melanin disruptor. In the formative stage, all feathers have interruption of the melanization process, in both eumelanin and pheomelanin. This then allows both forms of melanin to be barred. From this, such varieties as "crele" (barred red duckwing), barred buff or barred red emerge, in addition to the typical "barred black", which is the classic barred as in Barred Plymouth Rock or Dominique.

What we do know is that barring is a sex-linked dominant. Thus a homozygous B/B male when bred to a b+/b+ bird, will create barred heterozygotes (B/b+) in both sons and daughters. However, the female can never be homozygous for b, due to B being on the z-chromosome, thus she can only pass B on to her sons when bred to the b+/b+ male.

This gene seems to be heavily modified and f1 outcrosses and later generations often show a great many solid feathers, showing no barring at all, especially in the tail. Barring also shows an interesting interaction pattern. B produces both "barring" which is the very distinct black and white barring seen in show type BPR, and "cuckoo", which is a muddy, diffuse and indistinct form of barring seen in a great many varieties. The differences are due to other genes interacting with B. Two genes which have been put forth in this are Co and K. Co is Columbian while K is slow feathering. It is possible that factors in addition to these two are involved in this highly refined and distinct barring, as in BPR.

3. mo - mottling - This gene is an autosomal recessive. It creates a white feather tip with a black bar behind the white spot on pheomelanic feathers, but the black bar is not visible on eumelanic birds even though it is there.

Mottling is listed in the literature as an eumelanin diluter, but I have witnessed instances of mottling occurring on dark shanked birds. It is possible that there are several alleles of mottling, each behaving slightly differently. In some genetic backgrounds, though mo is recessive, there would appear to be some instances of mo behaving as a dominant. This could represent different mutations or differing levels of penetrance. While no clear research has been done on these aspects of mottling, they do occur and can often skew the "facts" of current published theory.

COMPOSITE COMBINATIONS

In this section we will look at how a variety is built. I have outlined above how the melanins (eumelanin and two forms of pheomelanin) combine to make all the colors in poultry. Then I established that the e-allele is responsible for the distribution and blending of the above melanins. Finally, I listed the genes which modify melanins in tone, distribution and pattern. This then gives us the tools to start assembling the genomes and genetically correct, universal names for the varieties.

There are five major varietal groups, which are based on the e-alleles. Each varietal group will have the same e-allele, and thus are "relatives" or variations on a theme, regardless of the "breed" they are on. By using the e-allele as the basis of naming and varietal systems, we can have a greater understanding of the relation of the varieties and greater ease in communicating.

In evaluating a given genotype, it is important to consider each gene I listed above. In this way, a bird can be given a total evaluation. While some factors may only reveal themselves through time and test matings, many are detectible in the phenotype. I use the chart below for documenting my own breedings. It is efficient and easily used, though it may take a few years to fill them in as testmatings are done and heterozygous recessives found or easily confused genes like Co and Db are worked out. In instances where I know a given gene is present and is homozygous, that is marked ++, while gene that is present and heterozygous is marked +-, while a gene which is known to be absent is marked --. If a given genes presence is not known, then that slot is left blank until one of the other three states is verified. The e-allele is noted, along with it's name and this then defines the variety group which the variety falls into. If the e-allele is heterozygous, this should be noted also, and will indicate a bird which can be used in breeding within two varietal groups, but it is not a member of either particular varietal group itself. Note both e-allele varietal groups in such a bird's genomic chart. Again, this may take some time to confirm.

Here then is the chart

E-allele
Ap
ap+
s+
S
Di
di+
Cb
ig
I
I-D
Bl
“rc”
lav
Mh
Co
Db
Ml
ch
“rb”
c
Pg
B
mo

Next, let's look at how a variety is mapped out on this chart. I will use wild type as our example.

E-allele - e+ - duckwing
Ap - ++
ap+ - --
s+ - ++
S - --
Di - --
Cb - --
ig - --
I - --
I-D - --
Bl - --
"rc" - --
lav - --
Mh - --
Co - --
Db - --
Ml - --
Ch - --
"rb" - --
c - --
Pg - --
B - --
mo - --

Now I take the genes that are present and write them out starting with the e-allele and listing each gene then in the above order. For this example: e+/e+ Ap/Ap s+/s+. This is the shorthand version of the full chart and notates what is present. When this is then converted into a varietal name, the genes present are listed in reverse order with the e-allele (varietal group) last. Here is an example of the above; Red (s+ and Ap) Duckwing (e+). Thus red duckwing is the varietal name of wild type, and the red junglefowl is red duckwing. This then is the system by which the varietal names are determined. Thus nothing is arbitrary and all the names in this system are in the same order and utilize the same name for each gene, so that genomic relation is easily understood. This then can allow for communication and consistency never before known in poultry.

One important last point is that certain interaction groups are named while their specific genome is not mentioned in the varietal name. Some good examples are Pg recombinants and s-allele-Ap/ap+ recombinants. These will be noted in the below varietal groups.

Now that I have outlined how this system works, let us look at the varietal groups which are the major e-alleles commonly found in the many poultry varieties.

All of the varietal groups are based on the e-alleles, thus the five varietal groups are as follows:

1. Duckwing (e+ - wild type)
2. Wheaten (eWh)
3. Brown (eb)
4. Birchen (ER)
5. Extended Black (E)

Throughout the wide range of breeds, the forms occurring amongst these five varietal groups are known by a bewildering array of names, many rather old and antiquated, which do not help to foster communication between poultry breeds around the world. Even more surprisingly, identical varieties in the same country, but occurring on different breeds, may have completely different names, and in some instances, these are the same name used in a different breed to describe a completely different gene combination (variety). Examples which come immediately to mind are "partridge" and "dark". I thus propose this new genetically focused naming system to create a consistency that allows common names amongst all forms of poultry and fosters much greater levels of communication (and most importantly) understanding, amongst all poultry keepers. I believe in this time, when so much of our genetic diversity has degraded and we have the ability to communicate globally, we owe it to ourselves, our birds and our descendants to foster communication and understanding. In this way, we can make much greater strides in maintaining our domestic fowl diversity globally.

To do this, I believe it is most important to understand that all chickens are ONE species. The breeds are NOT separate species. Species have been separated through great spans of time and can not produce fertile offspring when mated. All our poultry make fertile offspring when bred to each other. While I love the phenotypes of the various breeds (all of them actually), I believe we must focus away from tightly bottle-necked "pure breeds", and start to think in terms of wider relationships. Ancestry, relation, fowl type, desired traits, these all bear the greatest relevance. Selection for strength, hardiness, survivability, reproducibility, and genetic integrity, as well as the elimination of deleterious factors are paramount at this point in time, if we are to maintain fowl into the future. The varietal system based on the e-alleles is a base for working with color and pattern. While this is less important than the true integrity of the fowl, it is the starting point in breeding for color.

It is important to understand the e-alleles. At this point, if you do not feel confident in your understanding of the e-alleles, please go back to that section and study it in greater depth, then come to this section. Without a full understanding of the e-allele, you will not grasp the importance of the e-allele mutations in each variety. There are only a tiny handful of varieties that can occur on more than one of the e-alleles. For the vast majority of them, if you do not have the right allele, your "variety" will not be "proper" as to the standard description. My major focus in this book will be the "standardized" color and pattern forms. However, that is certainly not the end of the story, as only a handful of possible color varieties are "standard". I encourage the creative to consider "non-standard" varieties as there are many out there already and many more that could be achieved. Use the formula for naming I have outlined above for these non-standard varieties and bring them into the greater range of genetically correct naming which will be much more vast than the "standard" varieties.

Further another odd point with many standard descriptions is that they call for heterozygotes in many notable examples and a great many of them actually call for double-mating to achieve the male and the female as per the standard in question. Where this is the case, I will note it as so and list both separately. In developing new varieties, I would eschew double-mating schemes and focus toward homozygosity of the genes most important in creating a reproducible and consistent phenotype.

The Varietal Groups

Duckwing - (e+) - This is the wild type allele, and has thus been with us from the start. It should be no surprise then that there are many varieties on this e-allele. The greatest number of duckwing varieties are seen on the ancient game strains. This allele is so associated with games in fact that the varieties on e+ as a group are often referred to as "game colored". There are some notable examples that are not games which occur on e+. One that comes to mind is the "quail" pattern as expressed on D'Uccle.

The defining point of this allele is the female, as her back and her breast are different. The breast is pheomelanic while the back is pheomelanic and eumelanic, usually showing stippling, though pattern gene and melanotic together can result in double or triple lacing. Regardless of the exact shades seen, there will be a distinct separation of the back and the breast. While some eWh/e+ hens will show a division of breast and back that is similar to the e+/e+ hen, and certain eWh/eWh hens have a division that is entirely in pheomelanin, the presence of eWh will be clear in test matings through the eWh chick down which is very distinct. However, highly modified recombinant e+ varieties can be obscure or difficult to assess visually, and may require the assessing of greater evidence; i.e., adult hen patterning, the genes present and their actions, and chick down in test matings.

In males, hackle striping can vary greatly, depending on the genes present. All the melanizers will increase hackle striping. The hens are generally all hackle striped, however certain combinations can produce a clean hackle (red shoulder yokohama) where the stripes of the hackles are obscured rather than absent.

Wheaten - eWh - Wheaten is a mutation toward restriction of eumelanin and extension of pheomelanin. Thus the most commonly seen varieties on this allele, in addition to red wheaten (eWh/eWh s+/s+), are pheomelanically extended, such as black tailed buff, black tailed silver (black tailed "white"), black tailed red, to buff and solid red. It is the reduction of eumelanin and the extension of both sex linked and autosomal pheomelanin that allows these phenotypes to emerge. In my tests, I have found it impossible to create a solid black bird on this allele, and none of the eumelanizers are as effective on this allele as on other alleles. Wheaten also become "recessive" in the presence of eumelanizers to both e+ and eb, while in the absence of eumelanizers, it is dominant to these alleles. Thus eWh has a very interesting relation to eumelanizers in general.

The chick down for almost all varieties of eWh range from snow white, to yellow to pale orange with no patterning. In some rare recombinants with eumelanizers, some slight striping can be seen on this allele's chick down (Redcaps for instance). Solid white can be made on this allele by combining S/S with ap+/ap+ and I/I.

While hackle striping can occur in some hens, it is very rare in males and indicates the presence of one of the eumelanizers being present. Many claim that males of eWh are naturally a lighter red in the hackle than are e+ males, yet this is almost certainly the result of Di rather than the allele eWh itself.

In many instances, eWh makes a lighter underfluff on adult

birds, but this is not a hard and fast rule, as eumelanization can produce darker underfluff on this allele.

Brown - eb - Brown is an eumelanically extended allele. Thus "darker" forms are more common on eb. The most eumelanically restricted forms seen on this allele are exemplified by the penciled Hamburg varieties. This allele produces dark hackle striping in both males and females, so this is most often seen in this allele's varieties, and the major exception is when Db (ginger) is present in the absence of eumelanizers (as in the penciled Hamburg forms). There are a wide number of varieties on this allele and many of them utilize various combinations of Pattern gene to great effect.

I have found in test matings that due to the eumelanin extension of this allele, the creation of fully pheomelanically extended forms, such as buff or self red are very difficult on this allele when homozygous. One very interesting point of this allele is that self black fowl can easily be made on it with the addition of Ml and/or the "recessive black" type genes. This allele is responsible for "yellow legged black" varieties.

Chick down can vary wildly on this allele, depending on the other genes present. It is difficult to determine this allele absolutely by chick down alone.

The hens are much like the e+ hen, except they do not have the "salmon breast" caused by Autosomal Pheomelanin. The entire body of the eb hen will be the same "tone", color, or combination of tones. In unmodified eb hens, the body is similar to the back of the e+ hen, but somewhat darker.

While the male of eb, e+ and eWh are very similar, the hackle striping of eb is generally so distinct as to set it apart from the others, even when they are melanized.

Birchen - ER - This allele is also an eumelanin extension, but is more closely aligned to black in it's basic form than is eb. ER is basically a solid black bird with some areas of pheomelanin. In the male, the hackle, saddle and shoulder are pheomelanin, along with some lacing of breast feathers. The hen is the same, being a black bird with pheomelanic hackles and breast lacing in pheomelanin. However, oddly enough, ER allows for more complete extension of pheomelanin than does eb. Some well known examples of this are the combinations of Db and Pg, such as silver spangled Hamburg (which has a solid silver tail with black spangles), Seabright and polish (with pheomelanic tails with a black lace at the edges) and the Campine which has fully patterned tail feathering. This is the major point which defines pattern gene varieties on eb from those on ER; i.e., extension of pheomelanin into the main tail. Thus, it is highly likely that fully pheomelanically extended forms such as buff or self red, could be achieved on ER.

Self black birds are easily made on this allele through the addition of eumelanizers. Such black birds are notable as chicks by being solid black to black with a bit of reddish brown on the head. This sets them apart from E black birds, which are marbled with yellow to white and black as chicks.

In most instances, ER presents a chick down which is black with a reddish head, even when S is present. In advanced recombinants chick down may vary and can be confused with certain recombinants seen on eb, (Seabright chicks as compared to light Brahma chicks).

Hackle striping is usually present and heavy in these varieties in both sexes, even in the presence of Db and this may be the one factor that would keep a self buff or red bird from being achieved. More work needs to be done to determine this.

While several varieties are seen on this allele, it is rare in the hobby overall and under-utilized, especially when one considers the many lovely combinations that can be made on it. In the hobby, only the silver version on ER are called "birchen", yet the allele is named birchen by the genetic researchers, and thus I use the term birchen in naming all these varieties.

Extended Black - E - This allele is most common on self black varieties and their variants (blue, lavender, dun, barred etc.). While many think that E alone makes a black bird, this is not the case, and when unmodified, the E fowl looks much like the ER fowl, except there is no breast lacing and the hackle striping tends to be quite wide and dark. When eumelanizers are added, the self black forms are achieved.

The chick down of this allele is black and marbled on the undersides with yellow to white.

There are not any varieties which can be made on E utilizing Pg, as Db and Co have no effect on this allele. Thus the patterns seen on this allele are limited to barring (B) and mottling (mo). The vast majority of varieties on this allele are made by diluting eumelanin to other shades from blue to brown to white.

Perhaps the most complex variety seen on this allele is the Barred Plymouth Rock, which utilizes several genes to achieve the intense barring seen, yet they are not directly visible in the phenotype. Notable amongst these is Columbian (Co).

Now, before we get into the actual varieties, I want to mention a few terms I employ in the naming system I am outlining. There are a few instances where to list all the genes would be highly cumbersome. In these instances, I have used terms familiar to the hobby to describe these complex gene interactions. Where these occur in the varieties described, I will note it, but here I have to outline one variety to give you an idea of what I am talking about. Here we will look at the variety called "golden Seabright".

The genome of this variety is ER/ER s+/s+ Ap/Ap Mh/Mh Co/Co Db/Db Pg/Pg Ml/Ml Hf/Hf. To refer to all these genes makes the following long and complicated name; Hen feathered Melanized Pattern gene Ginger Columbian Mahogany red Birchen. However, to allow for some shortening, I refer to a variety like this as a Hen feathered Laced Red Birchen. This then tells us several things immediately. First, the Hen feathering is noted, and this gene modifies how the lacing is distributed in the male, making it like the hen (this is THE difference between Seabright and laced Polish pattern-wise). Second, we say "laced" and this immediately implies CoPgMl. Third, Birchen tells us immediately that the tail will be laced (as in Seabright and laced Polish which are both ER) as ER allows extension into the tail when Co and Db are present. As well, Birchen immediately tells us that Db is present, even in laced birds (which are typically CoPgMl) as Co by itself on ER does not allow for full pattern expression (as it does on eb) and thus Db MUST be present to have lacing on Birchen. It is only in such complex patterns that the names will be shortened in such a fashion and then it will be done not in an arbitrary manner, but in a manner which reflects the genetic reality of the recombinant.

COLOR VARIETIES AND PATTERNS

In this section I will endeavor to list all the commonly know varieties of standard poultry. I will not be looking at "non-standard" color patterns, but rather only at those which occur in at least one standard breed as a "variety".

One point which will be surprising to many people is that a good many varieties, as described by the standard, actually call for heterozygotes or have males and females that do not match genetically. This then requires the so-called double mating schemes, which in reality means you have two varieties, one for each sex.

Duckwing - e+

There are many standard varieties on e+. We will begin with the most basic form and work our way from there.

e+ Varieties Based on s+

Black Breasted Red, Partridge (in UK and parts of

Europe for games) - This is basically wild type. While there is some variation from breed to breed, for the most part, this variety is e+/e+ s+/s+ Ap/Ap. Hackle striping may vary from breed to breed with some being non-striped, others fully striped and yet others partially striped. The genetics of hackle striping has never been described in the literature, but it would seem this factor is controlled by two co-dominant genes which are yet to be described. The genetically correct name for this bird is **Red Duckwing**. BB Red as described by APA for the modern game is actually a "light brown" or dilute red duckwing (see below).

Light Brown - This is wild type which is Diluted. The genotype is e+/e+ s+/s+ Ap/Ap Di/Di. The genetically correct name is **Dilute Red Duckwing**.

Cream Light Brown - This is light brown with homozygosity for the recessive cream gene ig (inhibitor of gold). It is e+/e+ s+/s+ Ap/Ap Di/Di ig/ig. The correct name is **Cream Red Duckwing**.

Dark Brown - This is wild type with no Dilute and Pheomelanic Intensification from Mahogany. The genotype is e+/e+ s+/s+ Ap/Ap Mh/Mh. The genetically correct name is **Mahogany Red Duckwing**. Some extremely dark forms of this variety also incorporate eumelanin extension to further darken the red. These would be referred to as **Dark Red Duckwing,** as I use dark to refer to the extreme blackish red phenotype made from the combination of melanization with Mahogany. Such birds appear nearly black.

Blue Red - This is wild type to which one dose of blue is added. The genotype is e+/e+ s+/s+ Ap/Ap Bl/bl+. The genetically correct name is **Blue Red Duckwing**. When the Blue gene is homozygous, you get a variety called "splash red", though not standard, it is often seen from blue breedings. It is e+/e+ s+/s+ Ap/Ap Bl/Bl and is called **Splash Red Duckwing**.

Blue Light Brown - This is the diluted wild type with one dose of the blue gene. E+/e+ s+/s+ Ap/Ap Di/Di Bl/bl+. The genetically correct name is **Blue Dilute Red Duckwing**. The homozygote for blue is **Splash Dilute Red Duckwing**.

Blue Cream Light Brown - This is cream light brown with the blue gene added. It is e+/e+ s+/s+ Ap/Ap Di/Di ig/ig Bl/bl+. This is genetically **Blue Cream Red Duckwing**. The homozygote is Bl/Bl at the Bl allele and is a **Splash Cream Red Duckwing**.

Red Pyle - This is theoretically Dominant white added to red duckwing, however, in practice, these vary wildly. Ideally these would be e+/e+ s+/s+ Ap/Ap I/I, though the addition of Mh/Mh makes for a much richer tone of red on such birds. The presence of Di is often seen, but should be avoided, as it lightens the red, making it washed out and pale yellowish. The correct name for this variety is **Dominant White Red Duckwing**.

Crele - This is red duckwing that has barring added. In this variety, as for "red pyle", wide variation is seen. All the factors discussed for red pyle also applies to crele, and Di should be avoided, while Mh can enhance the depth of the red coloring. This genome is e+/e+ s+/s+ Ap/Ap B/B. The correct name is **Barred Red Duckwing**.

Ginger Red - This is red duckwing with Dark Brown (ginger) added. However, I have seen examples (especially in Old English Game) where this variety has been made using Columbian. However, the effect is of a much paler bird and is not the correct shade as described by the standard for this variety. The correct genome is e+/e+ s+/s+ Ap/Ap Db/Db. The correct name is **Ginger Red Duckwing**.

Red Quill - This is a rarely seen variety and is not a standard variety in the US, but is in other countries. This is the result of Pattern gene and Dark Brown (autosomal barring) on red duckwing. In other words, it is a patterned ginger red. The genome is e+/e+ s+/s+ Ap/Ap Pg/Pg Db/Db. The correct name is **Autosomal Barred Red Duckwing**.

Brassy Back - this is a dilute red duckwing to which melanizers have been added. To get a sound brassy back, you should have homozygosity for Ml and the "rb" factors. These vary a lot in practice though and variations of these two factors are seen frequently, with the well bred homozygote being seen rarely. The genome for this variety is e+/e+ s+/s+ Ap/Ap Di/Di Ml/Ml "rb"/"rb". The correct name is **Melanized Dilute Red Duckwing**. These are seen from time to time which are not diluted, but the standard calls for the lighter tones of pheomelanin which implies Dilute Red.

Quail - This variety is a brassy back to which Columbian is added. As per brassy back, these vary widely, and especially in the concentration of the melanizers. The genome is e+/e+ s+/s+ Di/Di Co/Co Ml/Ml "rb"/"rb". The correct name is **Melanized Columbian Red Duckwing.** In this example, to shorten the full name, the term "dilute" is dropped, as Columbian is a diluter of pheomelanin as well as an extender, and thus being Columbian on s+, they are thus Diluted. However, in testmatings, I have confirmed that these are in fact carrying the Di gene.

"Spangled" (as in Old English Game) **Speckled** (as in Sussex)- Not to be confused with true spangling as in Hamburg (PgDbMl), this variety is actually made with mottling. This is a fairly complex variety amongst the duckwing forms. The very dark coloring is made by combining Mahogany with the "recessive black" factors. The genome is e+/e+ s+/s+ Ap/Ap Mh/Mh "rb"/"rb" mo/mo. The correct name would be Mottled Melanic Mahogany Red Duckwing, but for convenience, the shorter form, **Mottled Dark Red Duckwing.** While "dark" is used for several varieties in the hobby, I use it ONLY to refer to the combination of Mahogany with melanization, which produces very dark reddish black phenotypes, such as those seen in the "spangled" varieties or the very dark exhibition type Rhode Island Red.

e+ Varieties Based on S

Silver Duckwing - This variety is one where the standard calls for male and female which do not match genetically. The true silver duckwing, which is a clean "white" silver can not have any Autosomal pheomelanin present. This is how the male is described in the various standards. This genotype is e+/e+ S/S ap+/ap+. This female however has a near white breast, unlike the standard description for a "salmon breast". Thus the true silver duckwing hen is considered a "male breeding-line hen". The standard description for the silver duckwing hen calls for the following genome; e+/e+ S Ap/ap+. However, the male that is this genome will be "brassy", i.e., not a true clean white, but creamy. These males are "female breeding-line males". The correct name in this instance is the same name used in the hobby; a very rare occurrence. In the APA standard, in the breed Phoenix, this variety is referred to as "silver", but it too is a Silver Duckwing, though it is called something different. Yet another good example of the arbitrary and colloquial nature of the variety names within the various breeds. It is very common to find Di in silver birds as well, though it is not confirmed that all lines have this gene. Hackle striping varies within these varieties, as with the red duckwing as described above.

Golden Duckwing - This variety is the combination of Silver (S) and Autosomal pheomelanin (Ap). The standard calls for a male which is a heterozygote and a female which is a homozygote. The female genotype is e+/e+ S/S Ap/Ap, while the male genotype is e+/e+ S/S Ap/ap+. As with the silver birds, Di is often present in this variety as well, though it varies and has little effect on the phenotype, this form being S and light in tone by it's very nature. As with both the Silver and Red Duckwing forms, the hackle striping varies. As with the silver duckwing, the golden duckwing is one of the few varietal names to reflect accuracy. Also, as with the Silver Duckwing variety of the Phoenix breed in APA standard, the Golden Duckwing form is referred to as "Golden".

Blue Silver Duckwing - this variety is identical to the

silver duckwing as described above. All the information concerning double mating would apply to this variety as well. The one difference with this variety is that it would have one dose of the blue gene, Bl/bl+. The name **Blue Silver Duckwing** is genetically accurate for this variety.

Blue Golden Duckwing - The above text pertaining to Golden Duckwing applies equally to this variety as well. Like the Blue Silver Duckwing, this is merely a Golden Duckwing to which one dose of blue is added (Bl/bl+). In both varieties (Blue Silver Duckwing and Blue Golden Duckwing), the homozygotes for blue are "splash" and make very pale color forms, with the Splash Silver Duckwing being nearly white. The name **Blue Golden Duckwing** is genetically correct for this variety.

White - It is not uncommon to find some self white varieties made on Duckwing, generally Silver Duckwing, as described for the "male breeding line". This form of white is seen with some frequency in white leghorns, which are fully clean Silver Duckwing (e+/e+ S/S ap+/ap+) with Dominant white added. In other instances, such genes as Columbian, Barring or Blue may be used in the mix to make a more fully white, a very clean white. Since this white on duckwing is not made in one particular way and can be made equally well with several different combinations, I will not list one particular way. These would be referred to as **Dominant White Silver Duckwing** or White for short.

Red Shoulder (as in Yokohama) - The red shoulder pattern as seen in the yokohama is the most advanced and complicated of the duckwing varieties. It uses multiple genes combined in a very layered manner to make the full phenotype. To begin with these are most often Dilute Red Duckwing, though I have segregated Silver from them as well, and thus it would seem that either of these genes will work for this phenotype. However, in most outcrosses the result was that the red shoulder used was Dilute Red Duckwing. Then on top of this, the PgDbMl (spangled - as in Hamburg) segregation group is added. Then Mahogany is added to achieve the intense dark red. Some birds also have the gene mo mottling, though this is not necessary to the final phenotype. Finally, the white is created not with Dominant white or homozygosity for Blue ("splash"), but with another diluter which is unique to the Red Shoulder in the standard breeds. This diluter has very odd segregation patterns and while it may mimic Blue and Dominant white to some extent phenotypically, it does not segregate in the fashion of those genes. I refer to this factor as RSY-D. It is dominant and the symbol is short for Red Shoulder Yokohama Diluter. So here then is the genotype list of the Red Shoulder; e+/e+ (S/S or s+/s+ or even S/s+) Ap/Ap Di/Di Mh/Mh Pg/Pg Db/Db Ml/Ml RSY-/RSY-D. This then would give us a very long name if we state every one of these genes. This is definitely a case where some shortening then is required. Thus, the name even when shortened would be White Spangled Mahogany Dilute Duckwing. Since this is still so long and cumbersome, I prefer to use **Red Shoulder Duckwing** (as in Yokohama) to identify this form. It is only in these very complex varieties where such shortening is acceptable.

Wheaten - eWh

There are several popular varieties on this allele.

eWh Varieties Based on s+

Wheaten - This is the hobby variety which is for the most part, the most simple form of wheaten, that is eWh s+. However, there is a lot of variation in how these are actually bred in the standard varieties. Some forms call for very pale, clean bodied hens and very light orange hackle/saddle in the males. These usually incorporate Di and ap+ to achieve this effect. There are darker strains which are eWh with Ap and di+. These are often seen in Malay and Aseel. Some of these strains also incorporate Mh to achieve a darker form, in which the shoulder is very dark red on the male and the female's back is particularly strongly colored. This too is seen in Malay, Shamo and Aseel, and while these forms are not as desired in the Old English Games, they are seen there too. The Cubalaya variety called "dark black breasted red" is also melanized and will be discussed further below.

Since there is so much variation in these birds, and all are considered the same "variety" - wheaten - it is difficult to classify them. Certainly the very pale forms seen in the OEG are eWh/eWh s+/s+ ap+/ap+ Di/Di. These would be called **Dilute Orange Wheaten** (ap+ is referred to as orange when in the presence of s+, as this is the color it makes; not red nor gold, but a true clean orange as is seen in the "wheaten" OEG). However, the darker birds as described for Malay or Aseel or Shamo but which are not extremely dark would be eWh/eWh s+/s+ Ap/Ap and if the hackles then tend to be greatly lighter than the hackles (but the hens body is not creamy and pale) there may also be Di/Di as well. I refer to all of these as **Red Wheaten**.

When Mh is obviously present due to the darker cinnamon/salmon color of the hen's back and the very dark red of the male's shoulder, the genome would be eWh/eWh s+/s+ Ap/Ap Mh/Mh and these I refer to as **Mahogany Red Wheaten**. Unfortunately, in the hobby, these are all just called "wheaten" or even "black breasted red". Since the birds do vary a good deal in the many breeds where "wheaten" or "black breasted red" on eWh is standard, there is no way to state one specific genotype for them, but from this discussion, you should be able to deduce what genes are present in your line of "wheaten" and be able to more accurately direct it in the way you want to go. As well, there is no one particular genetically correct name to encompass all of these. As listed above, each genotype would require it's own name as listed.

Dark Black Breasted Red (as in Cubalaya)- This is the darkest version of the above described wheaten variations. This form is most common in the Cubalaya, though admittedly, Cubalaya are seen in a range of segregations within this genotype. This form is also seen in Malay, Aseel, and Shamo. It is the above Mahogany Red Wheaten with melanization. The genotype is eWh/eWh s+/s+ Ap/Ap Mh/Mh Ml/Ml. Some examples may be seen where the recessive melanizers are present as well, and these further darken the bird, creating extremely dark hens and males and females with black at the top of the head. This however is not the standard form. The correct name for this form is **Dark Red Wheaten**, as when Mahogany and melanizers are combined to make extremely dark pheomelanin, this then is called "dark", as relates to red coloring.

Red Cap - The breed Red Cap only comes in one variety and this variety is based on wheaten. It is basically a Patterned Melanized Mahogany Wheaten. The crescent spangling on the female is very interesting in that Db is not present in this variety, which is clear by the black breast of the male. In the male, spangling is seen along the feathers of the wing bow and occasionally in some of the lesser sickle feathers. It appears that the combination of Pg Ml Mh are enough to make up this pattern. The old breeders called this color form "Crescent Spangled" based upon this pattern. This combination also creates an unusual aspect in the down, as it is an odd striping. This is created by the Pg Ml interacting with the down. The genome of this variety is eWh/eWh s+/s+ Ap/Ap Mh/Mh Pg/Pg Ml/Ml. The genetic name would be **Spangled Dark Red Wheaten**.

Black tailed Buff - This variety is made on wheaten and is a pheomelanically extended form, utilizing Co and/or Db to achieve this phenotype. Columbian makes a lighter bird that Dark Brown (ginger) does, when all other genes are the same. In the Black tailed Buff varieties, Dilute is usually present to create a nice, even, light shade of gold. Those made with Co are often too light and do not keep a nice soundness of tone. The birds made on Db tend to be of a better shade, with a darker top color with more depth. The genotype then is eWh/eWh s+/s+ Ap/Ap Di/Di Db/Db. The universal name would be **Dilute Ginger Red Wheaten** when made with Db, while those lighter birds made with Columbian are **Dilute Columbian Red Wheaten** and their genome would be eWh/eWh s+/s+ Ap/Ap Di/Di Co/Co.

Black tailed "Red" (as in New Hampshire) - This is a diluted bird, as evidenced by the over-all orange tone. This variety is a Mahogany version of the Black tailed Buff. The genome is identical to the Black tailed Buff except that the bird is homozygous for Mahogany. Also, many of these birds show evidence of melanization, in that the lower hackles show a stripe of black and the tail is very sound and solid black. The genome would be eWh/eWh s+/s+ Ap/Ap Di/Di Mh/Mh Db/Db (and in those with the hackle stripe) Ml/Ml. The universal name is Mahogany Dilute Ginger Red Wheaten though this can be shortened to **Mahogany Dilute Ginger Red Wheaten**.

Black tailed Red (as in Rhode Island Red) - These birds range in tone from the very dark exhibition type (I have a friend who jokingly calls these black tailed black) to a true red fowl with a black tail. These are both the same as the New Hampshire except they are not Diluted (di+/di+). The difference between the two forms seen in the Rhode Island Red is the recessive melanizers "rb". The true red form with the black tail is the same as New Hampshire with no dilution, while the extremely dark form is the same, except it also has homozygosity for the recessive melanizers. On wheaten, which is pheomelanically extended, when the Ginger gene (Db) is further extending the Pheomelanin, the recessive melanizers do not turn the bird fully black, but instead make a dark chestnut tone, which is near to black, but is not completely black. This creates the unusual near-black form seen in the exhibition RIR. The genome of the true black tailed red form is genomically eWh/eWh s+/s+ Ap/Ap Mh/Mh Db/Db, while the exhibition form is eWh/eWh s+/s+ Ap/Ap Mh/Mh Db/Db "rb"/"rb". The correct names would be **Mahogany Ginger Red Wheaten** for the true red form, while the extremely dark form is Melanized Mahogany Ginger Red Wheaten which can be shortened to **Dark Ginger Red Wheaten.**

Buff - This is the light version of the *full extension of pheomelanin*, with red (self red with no black tail) being the dark form. However, even in "Buff", there is considerable variation in shade and there are several variations which can be established as pure breeding. The genes making the basic phenotype are eWh, s+ Ap, Db, Mh, Di and Cb (which is the major factor in extending the pheomelanin into the main tail and main wing feathers). This combination when homozygous, by itself, can make the darkest form of true breeding buff. The darkest form of "Buff" is near to solid orange, being a yellowish orange color. The lighter versions are made through the addition of two other major genes; Co and I (Dominant white). When both of these are present on the above outlined genotype and are homozygous, the very light creamy buff colored birds, which are light yellow with an overall cream sheen, are created. In the lightest forms, those which are far too light for the standard description of Buff, Mahogany may well be absent too (mh+/mh+). Many buff lines will have only one or the other of these two extra lightening genes, while others are segregating both and show heterozygosity and variable phenotype. It is essential that all of these genes, in any of the combinations, be homozygous for consistency to be seen in breeding, regardless of the shade desired. The base genome for buff is eWh/eWh s+/s+ Ap/Ap Mh/Mh Db/Db Di/Di Cb/Cb. It is this combination of multiple pheomelanic extenders which causes the tail and main wing feathers to become fully pheomelanic. Once this is achieved, then the tone can be lightened with the addition of Co or I. The name for this variety is Pheomelanically Extended Dilute Red Wheaten, or **Buff Wheaten** for short. If the lightening genes Co and/or I are present, then the name would be **Light Buff Wheaten**.

Red - This is the dark form of full pheomelanic extension. It is similar to buff but is a darker, true red color. The genome is eWh/eWh s+/s+ Ap/Ap Mh/Mh Cb/Cb (without Di, this gene does not stop the true red color, but instead only helps to extend the pheomelanin into the tail and main wing feathers. This is basically the same as the Buff, except that it is not Diluted and is thus self red rather than self gold. The correct name for this form is Pheomelanically Extended Red Wheaten or **Self Red Wheaten**.

eWh Varieties Based on S

Silver Wheaten - This variety is uncommon in the US, but has become a standard variety of late in Europe. This is a fairly new variety and is not well set nor well understood. The full range of genes described above with the red wheaten varieties are seen segregating in the silver wheaten. The following genes are commonly seen; Ap, ap+, Mh, mh+, Di, di+. Depending on the segregations of these genes on the eWh/eWh S/S background, the birds may vary a good bit. Some will be brassy while others have various shades of red in the shoulder in males. A fully Silver Wheaten male should be eWh/eWh S/S ap+/ap+ mh+/mh+ and Di/Di can help as well. The hen should be the same to create pure breeding lines with clean males. The correct name is **Silver Wheaten**.

Salmon (as in Favorelle) - This is generally a silver wheaten, which tends toward Ap and Mh and sometimes di+. They are a darker form than one would want for true Silver Wheaten. The various lines of this breed vary widely. The standard seems to call for a heterozygote in the male as well, which makes a true breeding line impossible to achieve. The best lines I have seen, whether lighter or darker, tend to breed true and the males match the females. I will list now the genes in various lines, of which many seem to be segregating from line to line; eWh, S, Ap, ap+, Mh, mh+, Di/di+. The standard description of the male per APA would be eWh/eWh S/S Ap/ap+ Mh/mh+ Di/Di. The hen though seems to be described to allow for some variability, so that she is genomically the same as the rooster, but could be homozygous for any of the genes the rooster is heterozygous for. The correct name for this variety is **Mahogany Silver Wheaten**.

Black tailed White - This variety is best known from the Chabo, of which it is the most common variety. This variety is eWh/eWh S/S ap+/ap+ Db/Db. Like the black tailed Buff, these can occur with Columbian, though Db (ginger) is more common. Many of these also carry Dilute (Di), though the color form can be achieved without it. Those which have Ap instead of ap+ will have brassy bodies. The correct name is **Ginger Silver Wheaten**.

Brown - eb

There are many popular varieties on this allele.

eb Varieties Based on s+

Dark Brown (as in Leghorn) **Brown** (as in Cochin) - In the very darkest exhibition forms of the Dark Brown Leghorn, the birds are on eb, rather than e+. This is to allow for the extremely dark red coloring which is very difficult to produce on e+. It is easy to see that the eb allele is present in the hens, as the breast and the back are the same coloring, the hall mark of eb and something never seen in e+ hens. The Dark Brown utilizes melanizers to achieve the extremely dark red, in the same manner as it is done In the Rhode Island Red; Mh and melanizers. The genome of Dark Brown is eb/eb s+/s+ Ap/Ap Mh/Mh "rb"/"rb". The correct name is **Dark Red Brown**.

Partridge (as in Cochin, Wyandotte, Rock, Chanticleer) - This variety is Pattern Gene added to Red Brown in the most basic sense. In America, these birds are bred very dark and the genotype is eb/eb s+/s+ Ap/Ap Mh/Mh Pg/Pg. In England and Europe though, they are bred with Dilute in the genotype as well, making for a lighter bird. However the birds in the UK and EU are not all homozygotes and variation is seen in the Dilution and the Mahogany. The basic genotype for the lighter version would be eb/eb s+/s+ Ap/Ap Mh/Mh Di/Di Pg/Pg. Remember that the Di and Mh may be segregating. The genetically correct name for the darker American version is **Patterned Mahogany Red Brown**, while the lighter version as seen in Europe would be **Patterned Mahogany Dilute Red Brown**. Partridge is often double mated, with "male breeding lines" not having Pg and thus are actually Red Brown, rather than partridge. The "female breeding lines" have Pg and in fact, when this is well selected in the females, the males will show patterning to some extent in the body. The English have wonderful examples of such lines where the males are heavily patterned with double lacing.

Dark (as in Cornish) **Double Laced** (as in Barnevelder) - This is basically a melanized Partridge. The bird is a Patterned Red Brown to which is added Ml and "rb". The genome is eb/eb s+/s+ Ap/Ap Di/Di Mh/Mh Pg/Pg Ml/Ml "rb"/"rb". The correct name for this bird is very long, Melanized Patterned Mahogany Dilute Red Brown, thus we need to shorten this name. I have found a few varieties very difficult to give a shortened name that is still genetically telling, and this has been one of the most difficult. I have settled on **Double Laced Red Brown** for this variety. It is not as descriptive as it could be, but at least it is not six words long, though anyone wanting to be fully descriptive might consider using the longer versions.

Jubilee - This is the dominant white version of the Double Laced, though it generally does not have Di to enhance the shade of the red pigment, making is darker. It is not a standard variety in the US, where instead the white laced red takes it's place. The genome is eb/eb s+/s+ Ap/Ap Mh/Mh Pg/Pg Ml/Ml "rb"/"rb" I/I. The correct name, like the Double Laced is long and cumbersome; Dominant White Melanized Patterned Mahogany Red Brown. To shorten this one, I use **White Double Laced Red Brown**.

Buff Columbian, Buff (as in Brahma) - This variety is Columbian added to Red Brown. Columbian dilutes the red to make a buff golden tone. The lightest of this variety often also have Di, though it is not necessary to create the phenotype as described by the standard. The genotype is eb/eb s+/s+ Ap/Ap Co/Co. The genetically correct name is **Columbian Red Brown**.

Vorwerk or Belted (in Europe) - This is the melanized form of the buff Columbian and is identical to Lakenvelder except for the s-allele. The genotype is eb/eb s+/s+ Ap/Ap Co/Co Ml/Ml "rb"/"rb". The correct name is **Melanized Columbian Red Brown**.

Golden Laced (as in Cochin or Wyandotte) - This is the Columbian Red Brown with Pattern Gene and Melanotic added to achieve the lacing. The genotype is eb/eb s+/s+ Ap/Ap Di/Di Mh/Mh Co/Co Pg/Pg Ml/Ml. The correct name is Melanized Patterned Columbian Red Brown or **Laced Columbian Red Brown**. This form of lacing, on eb can easily be distinguished from those on ER Birchen, in that the tails of the eb based forms are black, while the tails of the ER based forms are laced in the tails.

Blue Laced Red - This variety is a darker version (no dilute) of Golden Laced to which one dose of blue is added to make the darker blue laced while two doses of blue are added (Bl/Bl) to make the light form of blue laced which is actually a splash laced red. The genome is eb/eb s+/s+ Ap/Ap Mh/Mh Co/Co Pg/Pg Ml/Ml Bl/bl+. The lighter colored or splash birds are the same genome except they are Bl/Bl. Some of these birds are segregating Db while "rb" is also seen in some lines. The correct name is alternately Blue Laced Mahogany Columbian Red Brown, or **Blue Laced Red Brown** for short, as Laced implies Co/Pg/Ml. The light version would be a **Splash Laced Red Brown**.

White Laced Red - This is the white version of the blue laced red. In other words, the blue eumelanin diluter has been replaced with dominant white, in most cases. A few strains of this variety are actually "splash" Bl/Bl homozygotes, but they are rare and the majority are based on I/I, Dominant white. The genotype is eb/eb s+/s+ Ap/Ap Mh/Mh Co/Co Pg/Pg Ml/Ml I/I. The correct name is White Laced Mahogany Columbian Red Brown or **White Laced Red Brown**.

Mille Fleur - This is similar to the Columbian Red Brown above, but is made using Db (ginger) instead of Co (Columbian). This can be told by the darker shade of red found on the mille fleur as compared to the lighter shade of buff gold seen in the Columbian variation. To this Ginger Red Brown is then added mottling to make the full phenotype. In some birds, Pg and Ml are present, but it is not essential to create the phenotype. The genome is eb/eb s+/s+ Ap/Ap Db /Db mo/mo. The correct name is **mottled Ginger Red Brown**. This variety is called "Porcelain" in Europe.

Porcelain - This is mille fleur to which the recessive gene lavender (lav) is added and is homozygous (lav/lav). This is **lavender mottled Ginger Red Brown**.

Blue Mille Fleur - This is mille fleur to which one dose of the Blue gene (Bl/bl+) is added. The genotype is eb/eb s+/s+ Ap/Ap Db/Db mo/mo Bl/bl+. The genetically proper name is **Blue mottled Ginger Red Brown**.

Golden Neck - This is a "pyle" version of mille fleur. There are two ways this variety is made; either by adding dominant white (I) to mille fleur or by having the "splash" version of the Blue mille fleur, which is a splash mille fleur. The version with dominant white is more proper, as the Dominant white makes for a more consistent and sound white in the eumelanic areas. The genotype is eb/eb s+/s+ Ap/Ap Db/Db mo/mo I/I. The genetic based name is **Dominant white mottled Ginger Red Brown.**

Golden Penciled (as in Hamburg) - This is a variety where the male and female descriptions as per the standard are actually describing two different varieties; one based on Pg/Pg and the other on pg+/pg+. The "male line" is pg+/pg+ which are basically ginger red brown. The complementary female shows no penciling. She too is a Ginger Red Brown. The "female line" is Pg/Pg. This male will show patterning in the body, with some being as well patterned as the females. The females show penciling throughout the body. Some lines are seen that are segregating Pg and in such lines a few of each type are produced, along with more heterozygotes; females with poor penciling and males with a few bars here and there in the body. For this reason, many people who have tried to breed this color form have found it to "not breed true" and never have discovered the secret; that there are actually two varieties masquerading as this one variety. The genotype of the male line is eb/eb s+/s+ Ap/Ap Db/Db (pg+/pg+). The correct name for this variety is **Ginger Red Brown**. The female line is eb/eb s+/s+ Ap/Ap Db/Db Pg/Pg. The correct name for this variety is **Patterned Ginger Red Brown**.

Yellow White Penciled (as in Friesian in the UK and Europe) - This is the same as Golden Penciled, except for the addition of Dominant white (I). The genotype is eb/eb s+/s+ Ap/Ap Db/Db Pg/Pg I/I. The genetically based name is **Dominant white Patterned Ginger Red Brown.**

Golden Spangled (as in Hamburg) - These fowl are the same as the female breeding line of Golden Penciled Hamburg, except for the addition of Ml. The genotype is eb/eb s+/s+ Ap/Ap Pg/Pg Db/Db Ml/Ml. Thus the Spangled is a Melanized Patterned Ginger Red Brown or a **Spangled Ginger Red Brown** for short. The spangled on brown eb can easily be distinguished from spangling on ER Birchen, as the eb spangled birds have solid black tails, while the ER spangled have pheomelanic tails with eumelanic patterning.

eb Varieties Based on S

Silver Penciled (as in Wyandotte, Rock, Cochin) **Dark** (as in Brahma) **Silver Partridge** (in Europe) - This is basically a silver version of the above described partridge variety, as bred in Europe and UK. The genotype is eb/eb S/S ap+/ap+ Pg/Pg. The correct name is **Patterned Silver Brown.** This variety is often double mated, as the "female line males" usually show considerable patterning in breast and body. The "male line" is basically a Silver Brown, with no patterning.

Columbian (as in Rock, Wyandotte, Cochin, etc) **Light** (as in Brahma) **Silver Columbian** (in Europe) **Ermine** (in Europe) - This is silver brown to which Columbian is added. The genome is eb/eb S/S ap+/ap+ Co/Co. The correct name is **Columbian Silver Brown.**

Lakenvelder or Belted (in Europe) - This variety is the

same as the Silver Columbian except that it is melanized. The genotype is eb/eb S/S ap+/ap+ Co/Co Ml/Ml "rb"/"rb". This variety has both of these melanizers to make the completely back hackles. The correct name is **Melanized Columbian Silver Brown**.

Silver Laced (as in Wyandotte, Cochin) - This variety is Silver brown with Co/Pg/Ml combined to make the pattern. The genome is eb/eb S/S ap+/ap+ Co/Co Ml/Ml Pg/Pg. The genetic name is **Laced Columbian Silver Brown**.

Silver Penciled (as in Hamburg) - This variety is one which is commonly double mated. In fact, the standard description for the male and female describes two different varieties. The male breeding line is not patterned, while the female line is patterned. This works in the same way as described for the Golden Penciled as described above. The genome of the male breeding line is eb/eb S/S ap+/ap+ Db/Db, while the genome of the female breeding line is eb/eb S/S ap+/ap+ Db/Db Pg/Pg. The correct name of the male line is **Ginger Silver Brown**, while the name for the female line is **Patterned Ginger Silver Brown**. It is common to see these lines which are creamy or brassy. This is due to Ap being present. Ap can not be present to get the clean "white" silver phenotype.

White - White can be made on eb using Dominant white (I). It is done either with S silver or with s+ which is covered by eumelanin. There are thus many ways that white can be made on eb with Dominant white. One of the most common ways to make white on eb is to add Dominant white to Silver Columbian. Since there is no one specific way to make this variety, I will not be listing a genotype or correct name. The most apt name is simply white.

Black (as in Wyandotte and Leghorn) - Self black can be made on eb, and is responsible for the yellow legged black varieties in Wyandotte and Leghorn. There can be any number of genes under the black plumage and either s-allele is possible. The key to this variety is homozygosity for both Ml and "rb". Since there is no one particular way to make these self black birds, I will not list a genotype nor a genetically correct name. A good way to describe such birds is as **Fully Melanized Brown**.

Birchen - ER

This allele has several very stunning varieties. However, while many varieties could be produced on it, few actually are. I would suggest those interested in developing new standard varieties on their breed might look to the varieties on this allele for inspiration.

ER Varieties Based on s+

Brown Red, Lemon Brown - This is the most basic of these varieties. The genotype is ER/ER s+/s+ Ap/Ap. However, amongst the various breeds, there is some variation seen in the descriptions of this variety of those breeds. In some breeds, the pheomelanin is described very light. In those birds Dilute (Di/Di) is also present. The Modern Game and Old English Game both require these Diluted versions, while the Cochin is not dilute in this variety. Thus, in Cochin, the brown red is correctly called a **Red Birchen**, while this variety in the OEG and Modern Game is correctly called **Dilute Red Birchen.**

Blue Red, Lemon Blue - This is the blue version of each of the above varieties. The genome is as above with the addition of Bl/bl+. The correct name would be the same as above, but with the addition of the word blue to each name; **Blue Red Birchen** and **Blue Dilute Red Birchen**.

Black tailed Buff - In some lines of OEG, the Black tailed Buff is made on ER. The genotype of these lines is ER/ER s+/s+ Ap/Ap Di/Di Db/Db. These lines are distinguishable from those on wheaten by the dark grayish underfluff of the ER based birds and by the very dark brownish reddish black chick down. The correct name for these birds is **Dilute Ginger Red Birchen**.

Golden Penciled (as in Campine, Braekel) - These birds are nearly identical to the version of this pattern on eb, except that the tail is patterned on ER and not on eb. The genome is ER/ER s+/s+ Ap/Ap Db/Db Pg/Pg. This combination creates down which is patterned, with broken striping and zig-zags on the back. The proper name is **Patterned Ginger Red Birchen**.

Golden Laced (as in Polish and Seabright) - These fowl are very similar to golden laced on eb, except that on ER, the tail is also patterned. The Seabright is further enhanced with both the male and female looking similar due to the presence of Hen Feathering (Hf). The genome is ER/ER s+/s+ Ap/Ap Db/Db Pg/Pg Ml/Ml Co/Co. The phenotype is best described as **Laced Ginger Red Birchen**. As laced implies

Co/Pg/Ml recombinant pattern, but on ER, Db is essential to the phenotype, so it is mentioned as well in the name, to fully express the phenotype. The down of this variety is golden with a small amount of Black around the head and underside.

Black - Some solid black fowl are made on ER by adding Ml and/or the "rb" type gene(s). These will be recognizable as being on ER by having solid black down or black down with a reddish brown head. This is distinct from E down. These would genetically be referred to as **Melanized Red Birchen**.

ER Varieties Based on S

Birchen - This is the basic silver (S) variety, being identical to the "brown red" (red birchen) except for the opposite mutation at the S allele and the Ap allele. The genotype is ER/ER S/S ap+/ap+. The proper name is **Silver Birchen**.

Silver Blue - This is the blue version of Silver Birchen. The genome is ER/ER S/S ap+/ap+ Bl/bl+. The Bl/Bl homozygote is called a Silver splash or simply a sport. The proper name for the Bl/bl+ heterozygote is **Blue Silver Birchen**. The name for the Bl/Bl homozygote is **Splash Silver Birchen**.

Silver Penciled (as in Campine or Braekel) - This is the silver version of the above Golden Penciled. The genome is ER/ER S/S ap+/ap+ Db/Db Pg/Pg. The proper name is **Patterned Ginger Silver Birchen.**

Silver Spangled (as in Silver Spangled Hamburg) - This stunning pattern is made on Birchen and obviously utilizes S. However, there are other genes there which help to produce the crisp white pattern, such as Di or ig (inhibitor of gold - cream). The basic genome is ER/ER S/S ap+/ap+ Db/Db Pg/Pg Ml/Ml, plus any of the other additive genes for further whitening in the Silver areas. It is the difference in the e-allele which makes the differences seen between the Golden Spangled Hamburg and the Silver Spangled. The GSH has a black tail, while the SSH has a silver tail with black spangled tips. It is the e-allele ER which allows this extension of pheomelanin into the tail. The proper name for this variety is **Spangled Ginger Silver Birchen**.

Silver Laced (as in Polish and Seabright) - This variety is identical to the above described Golden Laced (Laced Ginger Red Birchen). All the information there pertains to this variety also, except for the differences at the S and Ap alleles. The genome is ER/ER S/S ap+/ap+ Db/Db Pg/Pg Ml/Ml Co/Co, and in the Seabright, Hf (Hen Feathering) makes the male and female pattern identical. The genomic name for this variety is **Laced Ginger Silver Birchen**.

Black - Black can be made on Silver Birchen, just as it can on Red Birchen, by adding the melanizers. These would look identical in down to the Melanized Red Birchen, and it would only be through test matings that the S allele of this form can be determined. The genomic name would be **Melanized Silver Birchen**.

White (as in White Leghorn)- Self white can be made on Silver Birchen by adding either dominant white or recessive white, and further, by adding recessive white, a self white can be made on any variety I have here-in described, regardless of the e-allele. In the instance of Dominant white on Silver Birchen, these should be called Dominant white Silver Birchen, if this combination can be proven, however, in many instances, such as the White Leghorns, there can be many other genes there to "help" the dominant white fully whiten the plumage.

Extended Black - E

This allele does not have very many varieties, as it is very limited. For instance, Co does not have any effect on E, and Db has little. Pg likewise does not create any stunning patterns on this allele. However, Barring and mottling do show up in this allele so those are two varieties common to this allele and best displayed on it, while being less common (and somewhat less attractive) on the others. However, this allele is limited to self colors and the two mentioned patterns amongst the standard varieties. I will not divide this group by s allele, as they play little to no role on these varieties, as they are all fully melanized. Only Barred allows expression of the s-allele and that will be discussed in that varieties description.

Black - This is the most common variety on this allele. This variety can have either s-allele and many other genes can be present, though they will not have an effect on the visual phenotype. For instance, Co, Db, Mh, Pg, Ap or ap+ all may or may not be present under the black plumage of the E based self black fowl. It is highly likely that Ap is a contributing factor to the deep green sheen of some fowl, and Mh may play a part in this too. Conversely, ap+ is likely to have an effect on absence of green sheen and presence of purple sheen.

The genome for these black fowl can vary, but they will all be E/E at the e-allele and all have some of the "rb" (recessive black) group of genes. Some also have Ml to make the finished self black plumage. Since these can vary so much, I am not listing any particular genome beyond what I have described. The most correct name for this group would be **Melanized Extended Black**.

White - Another commonly seen variety on this allele, often utilizing Dominant white, but sometimes using recessive white, and in some instances, having both. The reason that white is so often seen on E is that the fully melanized black bird can easily be turned to white with the addition of Dominant white. Many genes may be found under these birds, and even S, Ap and Mh can be there, if they are fully melanized. As with the black, I will not list a specific genome, as it can vary greatly. The most accurate name is simply **Dominant white Extended Black** or **recessive white Extended Black,** as the case may be.

Mottled - this is the Melanized Extended Black to which the gene mottling (mo/mo) is added. While mottled occurs as a variety in several breeds, it is in the Ancona where it is best developed. In the Ancona which are very well patterned, Melanotic is often present, so it is possible that Ml helps to refine the mottling, making it more uniform and less diffuse as in other varieties, such as mottled Houdan. The most correct name for these varieties is **mottled Melanized**

Extended Black, or simply **mottled Black**.

Barred and/or Cuckoo - While these two forms are visually somewhat different, with Cuckoo being the more primitive and less defined of the two, they are very close to each other genetically, both being based on E, S and B.

While cuckoo is simply E/E S/S ap+/ap+ B/B (with the melanizers to make the bird fully black under the barring), Barred also utilizes all of these genes but also has Co and K (slow feathering) to make the very distinct bars. Further, there must be further modifiers at work, as even amongst barred strains, there are some with much more distinct barring than others, even when both are homozygous for Co and K.

S and ap+ is necessary in these varieties, to create a truly silvery-white bar. When s+ and/or Ap is present, these birds will be brassy or even look somewhat superficially like a "crele". As hens can only have one dose of B and one dose of K, they are darker and do not have as distinct a pattern as seen in males which are homozygotes for these genes.

The proper name for each type would be **Cuckoo Melanized Extended Black** and **Barred Melanized Extended Black**.

Blue and "Splash" - These two forms of this variety are the heterozygote and the homozygote for the Bl gene. These are made on Melanized Extended Black birds. The presence or absence of other factors, notably Pg, Co and the melanizers, can create various forms of this variety ranging from extremely well laced birds such as the best Blue Andalusians (Ml, Pg, Co) to non-laced blues, even in tone (pg+, ml+, co+). Further, the Bl gene can show wide variable expression and has a dosage effect, as seen in the "blue" forms (Bl/bl+) and the "splash" forms (Bl/Bl). Accurate names for these varieties would be **single dilute Blue Melanized Extended Black** and **double dilute Blue Melanized Extended Black.** While I have never worked with blue or splash that were on melanized ER or melanized eb, these probably exist somewhere.

Dun and Dun Splash - These are created in exactly the same manner as the blue forms, and all the information concerning blue applies equally to these dun and dun splash varieties, except that the gene I~D, which makes dun and is an allele of I (dominant white), creates a brown tone as opposed to the grey based tones of Bl. This gene segregates in much the same manner as Bl and show variable expression. Further, as with Bl, I~D shows interaction with the pattern genes and melanizers and also shows a dosage effect, as in Bl. The genetically based names would be simply **single dilute Dun Melanized Extended Black** and **double dilute Dun Melanized Extended Black.**

Lavender - This variety is made by using the recessive lavender gene, which is much like some forms of Bl in visual expression. However, lavender does create a different shade of gray-blue and also dilutes pheomelanin. However, in this variety, full melanization should be present and so no pheomelanin is seen. This then is a homozygote for lavender (lav/lav) on the Melanized Extended Black. The proper name would be **lavender Melanized Extended Black**.

Chocolate - This recessive gene is known in England and was described by Carefoot from his work with the gene. This gene produces a very similar effect to I~D/I~d+ (Dun), but is recessive. I have also seen and worked with birds from the

Serama breed in the US that have a similar eumelanic dilution factor visually similar to Dun, but recessive as is chocolate. This may well be the same gene, though no test-mating has been done to confirm this. This variety would be referred to as **chocolate Melanized Extended Black**.

The e-allele E has few varieties, all based on a "self" coloring of eumelanic extension. By combining the pattern genes B and mo with the various dilutions of the E based varieties, the whole range of varieties possible on E can be explored.

This then concludes the listing of standard varieties in their respective e-allele and s-allele groups. There are undoubtedly varieties which I have missed, at least in certain parts of the world. As well, there may varieties here which some are not familiar with and are not part of their standard, but are part of standards elsewhere. This certainly is not an exhaustive listing, but is comprehensive enough to illustrate the basic nomenclature and varietal structuring that most persons in the hobby will need to deal with. As well, I hope to have illustrated how common naming systems can help to create greater levels of communication between the various hobby groups. I want to stress that I am not asking you to stop using your colloquial names within hobby circles, where the terminology is current. What I am offering this system for is to further educated and productive communication within the various factions with the overall hobby.

Some of these names are bulky or longer than the traditional colloquial terms. They are not designed for shortening communication, but rather for expressing genetic facts when discussing varietal forms and thus allowing for understandable communication in all those cases where the colloquial terminology does not.

CONCLUSIONS

The most important thing which I have sought to convey herein is that all varieties are a complex structure of factors. They all begin with the e-allele, then the s-allele, and then from there, the many other factors (either present or absent) that modify these factors for shade of eumelanin and/or pheomelanin and pattern. All varieties can be understood in this way. By teaching yourself to think in this way about varieties, you can come to see all varieties on a continuum rather than as discrete units. Doing so can also foster greater understanding of poultry traits in general and thus allow the breeder to make more intelligent choices in outcrossing and to "think outside the box" when making those choices for outcross. It is very important that anyone, coming from any angle, learn to be a generalist concerning poultry genetics, as this allows you much great access and understanding.

One of the great failings of poultry breeding has been the notion that "breeds" and "varieties" are discreet units. That is, that they are different "species" or something very different and separate from each other. This is not the case. Breeds are collections of factors all in varying levels of heterozygosity/homozygosity which interact to make a visual representation or other trait. At most , they are regional variations showing high levels of homozygosity through long periods of isolation or through intense selection. At the least they are dominance interactions within highly heterozygous backgrounds which give an overall similar outline and variable but recognizable points of reference.

Clinging to these old notions has caused a great deal of genetic degradation in the modern breeds of poultry (both exhibition and commercial) through inbreeding and the single sire phenomena. This has occurred due to the pursuit of extremes in phenotype (visual or production) which has been fostered by the Mendelian understanding of genetics from the beginning of the 20th century and the concept of "purity". While interbreeding and "closed-line models" do foster high homozygosity of both the Dna and Rna factors within a line which produce the desired traits, it also magnifies the deleterious factors as well. This is especially true in strains, lines or breeds which incorporate deleterious factors through linkage and/or pleiotropy as part of their full phenotype (visual or production). Deleterious factors are more problematic than lethal factors. Lethals do not allow survival of homozygotes, while deleterious factors are often not lethal and thus can be perpetuated in lines for many, many years, eventually becoming so concentrated as to produce much greater effects. In some cases, these deleterious factors can reach levels which cause the extinction of lines through loss of viability. Many lines of poultry are reaching this level of concentration at this point in history and are faltering. Fear of outcrossing and improvement due to type or production concerns holds back the very thing which could save these lines. At this time, the wise breeder will become a generalist and work to restore viability and a low incidence of deleterious factors, and worry about bringing their type back to desired levels later.

It is my hope that this body of work can give the breeder a common reference point on color/pattern genetics so that they can maintain color breeding schemes while improving vigor, viability and strong reproductive ability. The method outlined here-in for variety construction can allow you to outcross in any direction needed and then have an understanding of how you get back to your color form. Further, I would stress that in an outcross project, it is not necessary to run back to homozygosity for color pattern or type immediately. The first few generations of selection should aim to actually improve the birds, getting out deleterious factors, raising disease resistance, increasing vigor and reproductive capacity while keeping the desired type genes segregating in the population. Once the bird is sound, then breed in large numbers to select for homozygotes for your desired traits that also incorporate the traits of your improvements. This is especially easy with dominant genes. With recessives, some homozygotes will be necessary, simply to insure the presence of the gene, but total homozygosity in the flock is not necessary.

To successfully improve lines, a "braiding" style of breeding is desired. This allows the blending of several lines into a recombinant which incorporates as many desired traits, from various lines, as necessary. The presumption of "braiding" is that you are going out to at least three breeds/lines, if not more. This is often necessary to bring together diverse and exceptional individuals which carry your desired traits and can pass them to their progeny. More lines may be necessary, especially if genetic diversity is desired from the start by using multiple males and females and blending their progeny into a cohesive mid-point, allowing for your desired phenotype along with the desired characteristics of a sound, healthy and productive line.

In seeking to have a broader overview of poultry genetics and strain building in general, I have formulated a list of trait groups in order of importance.

Traits Groups (in order of importance)

1. Vigor and Disease resistance - The most important traits. Without some level of vigor and resistance to commonly encountered pathogens, a line is doomed.

2. Fertility and production - A line which is loosing it's fertility is one which needs emergency work to correct this trait, which over time can lead to loss of viability and potentially extinction. Production is important first and foremost to ensure the numbers of individuals needed both to maintain genetic diversity and to allow for selection of traits. Further, production gives any line more value by adding to it's usefulness to man.

3. Temperament - Temperament varies widely, and many lines show high aggression toward man or each other, cannibalism such as feather picking, are very wild and flighty or easily frightened. All of these traits make fowl less enjoyable and more difficult to manage. Further, poor temperament creates stress for the fowl and the handler. A decline in production and/or the ability to keep males in sufficient numbers can all result in genetic bottle neck. Reducing aggression and flight response creates a calmer working environment for the keeper and the ability to keep more males. Both are very beneficial to keeping genetic diversity high through numbers produced and numbers kept.

4. Visual phenotype - This set of factors is perhaps the least important, for if you do not have #1 and #2, your line is in decline. #3 and #4 are the least important factors, though 3 outweighs 4, by increasing the numbers of males that can be kept (a very important and often overlooked source of genetic diversity - think "single sire phenomena") and by creating more manageable lines. It must be remembered that phenotype is window dressing and does not imply a sound or viable line. As long as you keep the proper genes segregating in your line, you can extract the visual phenotypes over time. The goal of a committed breeder would be to extract those desired traits on sound and viable birds, creating sound and viable lines, then the desired visual traits well set on sound and viable lines.

Finally, exceptional individuals are very important, as these can be founders in your improvement work and can also represent back cross material for improving the traits for which they are exceptional within your line, in later generations. An exceptional individual is one which has been tested and found to have good traits. During the first year, such birds are tested for resistance and vigor. Once breeding begins, birds are tested for their ability to pass on their good traits as well as their reproductive and (for females) production abilities. Into the second year, one tests these birds in many outcrosses to notate their genetic factors. If a bird has passed testing in the first year and is still being tested into the second, the potential is strong that the bird is exceptional for the traits he is strong in. Birds which have gone past their third year and are still being used and still rating high in their desired traits, may be ranked as "exceptional individuals". These birds must not show high incidences of deleterious factors, as one does not want to pass too much bad with the good if it can be helped. Remember the order of importance for the four traits groups above, and rank a bird based on these in every instance. Cull heavily and from the first for resistance/vigor and continue to watch for this throughout a bird's life. Breeding should show good fertility and viability, as well as the ability to pass traits. A well tempered bird, good in #1 and #2, which passes good temperament to some of it's offspring is an important individual. As always, give the least importance to visual phenotype and learn to notate the visual effect of heterozygosity for your desired type genes (where applicable for dominants). A bird that is good in #1 and #2, especially if #3 is good also, that is good in #4 is a very valuable bird indeed. When such a bird is from a "pure line" or shows high homozygosity for desired traits, it should be maintained as a founder individual and as backcross material for later generations. A good study of such birds is to keep them for many, many years, seeing how long they will naturally live, how long they remain fertile, how production drops over time and of course, to see if any deleterious factors appear later in life. In strain improvement, finding these exceptional individuals becomes paramount, as they can bring in greatly desired traits. Blending exceptional individuals is where the most progress can usually be made.

It is my wish that through the materials provided about color and pattern breeding in this book that many breeders can successfully improve their lines and keep the vast array of visual color and pattern phenotypes in existence for future generations. I can not stress strongly enough the need to look at poultry genetics from an overview perspective as one set of genetics, rather than only focusing on the narrow alleles seen in a given variety. Thus always remember the order of variety construction; e-allele, s-allele, modifier genes.

GRATITUDE

First, I would like to thank my grandfathers for encouraging my interest in fowl. They both encouraged me when I was very young and passed on to me a deep love for and interest in poultry.

Of all the researchers and authors, I must credit Fred Jeffrey first and foremost. Having straddled the worlds of the poultry hobby and the poultry sciences, Mr. Jeffrey demonstrated the interests of a generalist. Mr. Jeffrey was the first person to encourage me to look at all poultry genetics, rather than just those which applied to a certain breed. Mr. Jeffrey's contribution to the world of poultry has often been overlooked but he is my greatest inspiration. My deepest thanks.

Next I would like point out the work of Clive Carefoot. Here again we have someone who straddles the worlds of the hobby and science. My gratitude for his many contributions.

I have found the compendium ***Poultry Breeding and Genetics***, edited by Crawford to be an indispensable reference work.

It is from these sources that I have devised my own

testmatings and been encouraged to look further and (especially from Jeffrey and Carefoot) to think outside of the box.

Here is a list of books by the above three sources that I have found very valuable.

F. P. Jeffrey

1. ***Bantam Chickens*** (ABA Publications)
2. ***Chicken Diseases*** (ABA Publications)
3. ***Old English Game Bantams as Bred and Shown in the United States*** (F.P. Jeffrey and William Richardson)

Clive Carefoot

1. ***Creative Poultry Breeding*** (self published - UK)
2. Various published papers.

R.D. Crawford (editor)

1. ***Poultry Breeding and Genetics*** (Elsevier)

Next I would like to thank Dr. Ronald Okimoto for his contribution to the hobby through the many internet message boards which he has answered questions and passed information on. His has been an extremely valuable contribution and, I am afraid, an under recognized contribution. I would personally like to thank Dr. Okimoto for his personal correspondence with me and for his patience.

There are so many people within the hobby whom I would like to thank. Many people have shared stock and stories and data with me, that it would be impossible to thank them all.

There are a few however who stand out and must be thanked individually.

First, Danne Honour is a remarkable breeder and good friend. Danne has provided me with many, many old articles from many old publications on poultry. These articles have been invaluable in putting together an understanding of our poultry history throughout the last century and a half, as well as understanding the techniques and philosophies of the master breeders from the heyday of poultry showing and breed development and improvement. Thank you Danne!!

Next, Toni Astin and Brian Shamblin, who have been great friends and who have both shared stock and information with me freely. My best wishes to you both!

Grant Brereton has provided me with a lot of excellent information from the UK and Europe. Thanks Grant!

Barry Koffler was one of the first persons to have a really significant poultry website on the internet. His Feather Site (http://www.feather site.com) is an excellent resource, with a huge number of pictures of poultry breeds and varieties and an invaluable source to links for the internet poultry world. I have enjoyed Feather Site and find it very useful. I have spent many hours pouring over the collection of pictures. Thank you Barry!

Jerry Schexnayder has provided me with a large number of Serama for test matings. I am very grateful to have had this material to study. Thank you Jerry!

Finally, I would like to extend my thanks to the entire poultry community. Without the poultry keepers, breeders and researchers, there would be no materials to study! Thank you all!

My best wishes to you all.
Brian Reeder

NOTES

ABOUT THE AUTHOR

Brian Reeder is a well known writer of poultry articles in the United States. Having worked with poultry from his youngest years, he began serious research on Domestic Fowl genetics in his teens, setting up major test matings in his early twenties. These matings and collected data span a decade. Now, nearing his fourth decade of life, he is focusing on writing based upon his experiences and gathered data with the Domestic Fowl.

In addition to Domestic Fowl, Mr. Reeder also studies the genetics of Domestic Crucian Carp (goldfish), Colubrid snakes, and many other domestic animals and plants. For more information on his interests and to access his many articles, please visit his website at http://www.longtailfowl.com or http://www.panopliageneticus.com

www.ingramcontent.com/pod-product-compliance
Ingram Content Group UK Ltd.
Pitfield, Milton Keynes, MK11 3LW, UK
UKHW041926190726
13854UKWH00003B/1479